SERIOUSLY SILLY SCIENCE

A science reader for the whole year – and some of it is even true

by

Dr. Gary Loren McCallister
Professor Emeritus
Colorado Mesa University

Library of "Areyoukiddingme"

McCallister, Gary Loren
    Seriously Silly Science
    1-too-many edition
    Cheap version – no index or documentation

Executive Editor
    Did you really think anyone would admit to being the editor of this book?
Development Director: Obviously none
Editor in Chief: badly needed
Art Director: Nothing artistic about this book.
Creative director: You're joking, right?
Cover Design: Stolen from the internet

All rights are reserved – except no one seems to want them.

IT
COPYRIGHT 2015
All rights reserved
ISBN:1514239965

CONTENTS
- Introduction - 7
- January - 9
    - Week 1 - One being number one   10
    - Week 2 - Doing straight time   13
    - Week 3 - Tthe expansion of time   16
    - Week 4 - The time of your life   19
    - Week 5 - Caffeine   22
- February - 25
    - Week 1 - Getting ones wires crossed   26
    - Week 2 - Geek romance   29
    - Week 3 - Fundamental, inexplicable attractive forces   32
    - Week 4 - George Washington in poor humor   35
    - Week 5 - Presidential patent   38
- March - 41
    - Week 1 - A reasonable chapter about evolution   42
    - Week 2 - Science fair   45
    - Week 3 - In mourning   48
    - Week 4 - Decentralized systems   51
    - Week 5 - Husbandry   54
- April - 57
    - Week 1 - April fools green   58
    - Week 2 - The good, bad, and ugly   61
    - Week 3 - Living together   64
    - Week 4 - Counting things that aren't there   67
    - Week 5 - Ode to mediocrity   70
- May - 72
    - Week 1 - Mothers day science   74
    - Week 2 - River running   77
    - Week 3 - Drug evaluation testing   80
    - Week 4 - Cause and effect   83
    - Week 5 - Minimal measurement   86
- June - 89

- Week 1 - Bee venom — 90
- Week 2 - Po-entomology — 93
- Week 3 - The worth of insects — 96
- Week 4 - Art and science — 98
- Week 5 - Jell-O — 101

July - 104
- Week 1 - The science of freedom — 105
- Week 2 - Language and fireworks — 108
- Week 3 - Sunburn — 111
- Week 4 - Cool cat on a hot tin roof — 114
- Week 5 - A quart of light — 117

August - 120
- Week 1 - Hell — 121
- Week 2 - Rock skipping — 124
- Week 3 - The "dog days" of summer — 127
- Week 4 - That lucky old sun — 130
- Week 5 - Normal — 132

September - 135
- Week 1 - The science of the Grammys — 136
- Week 2 - Brain lobes — 139
- Week 3 - Insects and flowers — 142
- Week 4 - Sweet medicine — 144
- Week 5 - Which way is up? — 147

October - 150
- Week 1 - Daylight savings time — 151
- Week 2 - An eye for an eye — 154
- Week 3 - Moles — 157
- Week 4 - Class II medical devices — 160
- Week 5 - Suffering — 163

November - 166
- Week 1 - The science of Thanksgiving — 167
- Week 2 - Feathers — 169
- Week 3 - Depressed turkeys — 170
- Week 4 - Yawn — 173
- Week 5 - It's getting cold — 176

December - 182
- Week 1 - Hole-y snow, Batman — 183

Week 2 – The cost of the Holiday	186
Week 3 – Counter current Christmas	189
Week 4 – Christmas	191
Week 5 – Christian science	194
Epilogue – Neutral	197

DEDICATION

For all those people who have helped me become what I am today, but who don't want their names made public.
You know who you are.

ACKNOWLEDGEMENTS

I would like to acknowledge all those people who have helped in the making of this book, but most of them have asked to not be identified. My wife would just as soon not take any responsibility for my writing either, but it would be strange to not blame at least one other person for such a mediocre undertaking. So it's partly her fault.

WHAT'S SCIENCE GOT TO DO WITH IT?

Well, actually, science has got an awfully lot to do with just about everything, except the things it doesn't. You might be surprised at the things, and number of things, science has something to do with.

Science has a bad habit of taking itself very seriously. A lot of science writing seems breathless and urgent. Science seems to always be about the next big thing, and each report claims that it is going to lead to something even bigger within just a few years. I think a lot of science writing is simply an infomercial for the next grant.

Personally, I find the science most interesting which doesn't make anyone any money and was only done out of curiosity, or to help someone out with a little problem. I love it when someone finds a way to use termites to help make brick for building in a wood-barren, third-world area. Or figures out how to drill for water using a merry-go-round.

Almost as good is when I find out that scientists don't know so much after all. Like, no one seems to agree about why we yawn. Why do we sleep? Why do we fall in love? How can we fall in love and also fall out of love? Why does love sometime turn violent?

I was chastised once by a scientist who criticized a comment about "a pints a pound the world around". He was correct that scientifically that is not true, but he didn't know that the phrase actually refers to a time when Britain had a worldwide presence and a pint of British beer was a pound (a British money term) everywhere due to the uniformity of the British financial system. Scientist can be myopic about things other than science.

Finally, while I know it's a cheap shot, because it's so easy to do, but it's kind of fun to occasionally tease scientists a little about their quirks, short-comings, and even, sometimes, their poor reasoning and logic. I can do this because I am one of them, and I exemplify almost everything I make fun of.

Scientists seem to think if science doesn't have anything to do with something, that thing doesn't matter. Of course, that is simply not true. There are all kinds of things that aren't scientific that are terribly important; like freedom, love, loyalty, perseverance and the number five. In fact. Most scientists wouldn't have become scientists without many non-scientific attributes, characteristics, and environments.

So this book doesn't take science very seriously. Sometimes, in fact, I just make stuff up. I hope that is part of the fun - figuring out when I am lying. These essays are presented as an annual, weekly reader, arranged by subject matter as it is appropriate to the months of the year. All months have five essays. I know not all months spill over into five weeks, but I don't charge anything extra for the extras. Consider it a bargain.

JANUARY

Themes:
- Time
- New Beginnings
- Hurry
- First
- Did I Mention Time?

That may seem like a narrow focus for an entire month, but stop and think. It's January. Nothing happens in January. January is probably the longest, slowest, most-boring month of the year. It's cold, dark, and quiet. Most people either have seasonal affective disorder (SAD) or are hungover from the celebration of the New Year. It only makes sense that this month's essays deal with beginnings, time, and being in a hurry.

JANUARY

Week 1

ONE BEING NUMBER ONE

They say the first chapter ought to be about the first chapter by way of introduction. Being the number one seems kind of restrictive to me. I guess that means this is the first chapter of many chapters making it number one. The other chapters don't exist yet, of course. I am not sure how many there will end up being, maybe fifty two or so. But it doesn't really matter until you have one, the first one. So that is why I am writing about the number one.

Actually, it doesn't matter how many numbers of chapters I end up with. It will always be as many as I have previously written plus the last one. Let's say I wrote 5,721 chapters. I would just right the first one, then add the second, and then add the third, until I reach 5,721. If I write the chapters on a weekly basis, I would then be one hundred and seventy five years old. (You figure it out.) So the number one can be pretty interesting, even if my first chapter isn't.

But instead of adding chapters, it could be more interesting if I could multiply them. If I multiply one chapter by one chapter I just get one chapter. But if I multiply eleven chapters by eleven chapters, I get 121 chapters. And if I multiply 111 times 111 you get 12321; and 1111 times 1111 equals 1234321. Notice anything? These are like palindromes, words that spell the same front and back, except they are numbers. It goes on forever.

Another thing you can do with the number one is add it to zero. But $0 + 1 = 1$. However, if you add the last two numbers together, $1 + 1 = 2$. Then if you add the last two numbers together again you get $1 + 2 = 3$. This series is called the Fibonacci Series after Leonardo Fibonacci who lived one thousand years ago. He discovered that, if you continue this operation, you get a series of numbers that would look like this: 1, 2, 3, 5, 8, 13, 21, 34, 55, 89, 144 . . . . OK I'm bored.

It turns out that this kind of series of numbers shows up a lot in biology. For example, sunflower seeds are set in spirals to the right and left within the blossom. And the number of spirals is always one of the numbers in the Fibonacci series. This arrangement apparently keeps the seeds uniformly packed no matter how large the seed heads are. The number of spirals in pine cones is usually a Fibonacci number. Most flowers have petals that are a Fibonacci number.

It's not hard to bring science into mathematics, although some would say that mathematics has brought science into it. Because mathematics is the study of quantity, structure, space, and change, (that's what it says on Wikipedia, so it must be true) it has been a handy tool for the sciences. Scientists and mathematicians alike examine the world for patterns and try to draw conclusions from those patterns to predict events, or help us control our world.

Of course, there are always basic assumptions that are agreed to by all, and then reason is applied to those basic axioms, definitions and assumptions to arrive at what scientists and mathematicians hope is the truth. But it is best to remember that reason, for all of its value, is only as good as the basic assumptions on which it is based. This means that if you change the assumptions, the truth can change also. Sometimes scientists forget that their reasoning is based on their assumptions, and then they make an asses of themselves.

Well, I don't expect my chapters to multiply or follow the Fibonacci Series, but they may be related to mathematics in one last manner. Einstein once said about mathematics: "As far as the laws of mathematics refer to reality, they are not certain; and as far as they are certain, they do not refer to reality." Hmmmm. Thanks for reading my first chapter.

JANUARY

Week 2

DOING STRAIGHT TIME

Sometime in the 1700's people began to conceive of time as linear in measurement. This was unlike earlier times when time was considered circular. There have been times when time has been perceived as random with no pattern at all. Most of the time we say that time is uniform, but there are times when time does seem to go faster or slower. January seem like a good time to write something about time. I guess you could say that "It's about time!"

Contrary to what many believe, I am not old enough to recall ancient and primitive conditions. However, the passage of time has always brought about unpredictable and dangerous changes that often result in dissolution and death. I have experienced my share of dissolution, and I can affirm it is dangerous, and could even sometimes result in death.

For most of recorded history, though, humans have measured the passage of time in natural cycles that match the diurnal rotation of our planet, lunar months, and the rotating seasons. The measurement of cyclical time allowed greater organization of society and control over the elements. Humans learned to perform certain deeds, such as planting or hunting, at the "right times", and in the correct sequence. These are lessons we have somewhat forgotten in our modern, anything goes, world.

Cyclical time introduced a moral dimension to mans' thought as well. Things could be said to have been done at the wrong or right time. In addition, each generation could begin to compare its behavior to the behavior of its ancestor's generation during similar periods of time. People who lived under a cyclical paradigm valued patience, relatedness of parts, ritual, relationships, nature, the healing power of time, and the symbol of renewal and resurrection.

Thinking of time as a linear experience, with a distinct beginning leading to a unique ending, is nearly universal in modern thought. The concept was recognized by ancient Greeks who hoped that reason would improve mankind's lot by enabling him to avoid mistakes of the past. The Romans felt that time led people down a path that culminated in a glorious destiny. Or not, as the case may be.

However, it was the rise of the monotheistic religions that suggested that mankind's fate might be directional. It was in the sixteenth century that the inventions of science, combined with the Reformation, began to spread through Europe leading people to speculate about the origins and the end of the earth. This linear timeline is carried forward in today's science where we debate such things as the beginning of the universe and the beginning of life. As well we debate when the end of the universe will occur and when someone is actually dead.

The idea of linear time assumes that mankind is on a trajectory of progress. Men disagree about when time began and when it might end. The overall assumption, however, is that we are on a straight line of progress toward better things. This thought so pervades modern America that it has shaped our entire culture. In contrast to cyclical time, modern culture values haste, practicality, concentration, efficiency, analysis, direction, speed, power and control of nature.

Modern science seems to believe we are simply on an extrapolated timeline from the past. Perhaps, if one does not see any possibility of deviation from the trajectory in the future, they will consider any deviations in the past as insignificant. In fact, the past is assumed to have led to this moment when we are just lucky enough to be in existence at the exact apogee of human existence. Really?

One thing worries me. Straight lines do not always lead upward. They can just as easily be drawn downward. All measures of time that we use are based on measurement of repetitive activities, whether they are the vibrations of an atom, the rotation of the earth on its axis, the lunar cycles, or the seasons in a year. Maybe those circular comparisons should be enough to give us pause about modern times. It's about time.

JANUARY

Week 3

THE EXPANSION OF TIME

I recently had a birthday. My family made guesses as to my age, and their estimates were . . . interesting. As I considered some of their guesses, though, I began to notice an intriguing pattern. This led me to do some basic research, and I think I have discovered some fundamental relationships that underlie the data.

The earliest known record of my age is based upon my birth. However, for the life of me, I can't recall the moment or the date. I also discovered I don't have that particular certificate. This immediately begs the question of whether or not I could be elected President of the United States, so I have decided not to run.

The first official proof of my existence appears to be my graduation certificate from high school. Again, for the life of me, I don't really recall that event either. But if I assume I did not graduate early, a fairly safe assumption, and that I was also not held back a grade or two, a rather shaky assumption, then I should have been close to eighteen years of age at the time of that event.

Stay with me here. The science part is coming. I also found my marriage certificate which was dated some time later. It appears to have something on it that looks like a date of birth, but the whole thing is in German so I'm not sure. I do recall this event vividly. It was the major miracle of my life. Anyway, it appears that I may have been twenty one years old at that time.

My military discharge papers, surprisingly honorable, indicate that I was twenty two years old on discharge. That is if government records are to be trusted.

After that the trail gets sparse, and my present age is based upon these somewhat shaky assumptions. If this data applies to the general population, and if one graphs the data, they suggest a very real and fundamental relationship.

The physicists treat time as if it were a variable. However, I'm not sure they understand the full implications of doing that. If the estimates I made above are accepted as fact, or at least taken seriously within a margin of error, one cannot escape the following conclusion. As time goes on, the age of the average person not only increases, but actually does so at an accelerating rate.

Since this theory is based entirely on past data, we can conclude that age increases occur only in the past. If that is true, then the point in time representing a person's birth may be moving backward in time, but at an ever-decreasing rate. Whether aging occurs in the future is not certain. If, however, this pattern continues into the future, we can be sure that at some finite period of time, we each will become infinitely old.

There are serious and pragmatic consequences to these conclusions. First of all, all questions concerning my birth are permanently relegated to past history. Secondly, all speculation about life prior to my birth is equally meaningless.

More research is needed in this area, and not all questions have been cleared up. For example, perhaps this trend in aging is because of some fault in time itself. Where is the evidence that time moves at a constant speed? There is anecdotal evidence that time can seem slow and, at other periods, seem to speed up. These occasions have been dismissed as a result of such questionable properties as fun and despair. These subjects are not generally quantifiable and, therefore, not subject to serious scientific consideration.

Another possible explanation of the "time" thing is that universal expansion of space is accompanied by universal expansion of time. Or perhaps the space expansion is caused by time expansion. Or the time expansion . . . err, well, I kind of lost my train of thought there. Do you suppose you just wasted some time?

JANUARY

Week 4

THE TIME OF YOUR LIFE

"It was the best of times; it was the worst of times . . . ." No wait, that's already taken. How about, "To everything there is a season . . . ." What? That's already been used too? Then how about, "The time of your life"? I know it's a cliché expressing the quality of an experience, but I really mean, the measureable "time" of your life.

Our lifetimes take place at certain rhythms that are sometimes obscured and forgotten in our modern world because we are driven by technology. It might be interesting to consider how things happen through biological time.

One of the fastest processes that occurs in your body is the transmission of information along your neurons. The rate is variable depending on a lot of factors. In peripheral nerves from your big toe to your spinal cord, information can transfer as fast as 225 mph, or 330 ft/sec.

Some muscles work quickly and others work slowly. Your eyelid muscles are especially quick. They can contract in about 1/300th of a second. That is what makes a wink so deadly. A wink can happen so fast that you aren't really sure if the winker actually winked at the winkee or not. Then what is the winkee to do? If they wink back and the first winker hadn't really winked, then that could be embarrassing. If they don't wink back, smile, or something, they may miss an important opportunity.

But maybe winking is old fashioned. Does anyone wink anymore? Obviously not at me. However, be advised, if you are going to wink, and you really want it to have an effect, do it slowly and exaggeratedly. (See, science can be so practical.)

Emptying your stomach can take anywhere from fifteen minutes, if all you put in it was water, to up to six hours if you ate a big, greasy, pepperoni pizza. Yes, the pepperoni can still be there at 2:00 am. The time it takes, on average, for food to move entirely through your digestive track, from start to, shall we say, finish, is about 12 hours. If you hurry things along much faster than that, there isn't time for your body to remove enough water from the food. Then you will have what is politely called a loose stool.

Did you know that if you get less than eight hours of sleep, all your immune cells are measurably less effective? Certain cells that attack and engulf foreign invaders lose efficiency with the loss of only one hour of sleep. Effectiveness goes down until when you are only getting about six hours of sleep. Then it levels off. At that point you can rest easy that you immune system can't get any worse.

Other immune cells secrete antibodies when you get sick. Even if you are healthy and well rested, it can take up to two or three weeks to produce enough antibodies to be effective. That's why they say that, if you eat right and get plenty of rest, you can get well in two weeks. But if you don't, it's going to take fourteen days.

The truly magical number in human biological time is six weeks. That probably doesn't sound familiar to you, but consider these facts. Wound healing generally takes about six weeks. Obviously the timing depends on the wound, but even minor surgery requires about six weeks to repair the damaged artery bed. And guess how long it takes to remove all the nicotine from your body after you quit smoking? About six weeks. If you start a new training regimen, it will take about six weeks to grow new muscle. If you start learning a new skill, like playing a new song on a musical instrument, it will take about six weeks to master. It even takes about six weeks to metabolize five pounds of fat.

Six weeks is such a universal estimate of biological growth in humans that I have decided to name this "time frame" a new unit of time called a "hexachron". Though we coordinate our lives around a 24-hour clock, a seven-day week, a 30-day month, and a 52-week year, biological time is something else entirely. The biological "time" cycle is most often about six weeks, a hexachron.

How many hexachrons do you have left this year? How many hexachrons, prior to summer, do you need to get into swim-suit shape?

JANUARY

Week 5
(If needed, although I don't really know
anyone who needs an extra week of January.)

CAFFEINE

(Attention readers: To get the best effect from reading this article you should read as quickly as possible and jiggle one knee while reading.)

Quick! I'm in a hurry! What's the most commonly-consumed, mind-altering substance in the world? Nope. Nope. Nope. It's caffeine. Perhaps you hadn't thought of caffeine as a mind-altering substance. But, of course, it is. Why else would someone consume a normally very bitter-tasting compound, if there was not some desired result? But we'll get to that.

Unlike other drugs, which are scarce in the natural world and have to be laboriously and delicately extracted or created, caffeine literally grows on trees and bushes, some kinds of cacti, lily, holly and camellia. In fact, I found that there were more than one hundred species of plants that produce caffeine molecules in their seeds, leaves, or bark.

Hurry! Can you tell me how many species of plants manufacture nicotine or morphine? Just one each! Quite a contrast. I wonder what caffeine is for in the real world. Maybe it's just so bitter it keeps consumers away. Either that, or it makes consumers so hyper they become inefficient. I'm just saying.

Human association with caffeine began a very long time ago. Different plants containing caffeine were discovered in different geographic regions of the world, but the use of all of them by humans was wide spread. Today, tea and coffee are the most popular drinks in the world. More tea is consumed, but coffee is a close second. However, in the

US, soft drinks may surpass coffee in consumption. The "cola" in some soft drink names comes from the "kola" bean, found in Africa, which is used for caffeine extraction.

Caffeine is manufactured by plants from a precursor molecule called xanthine. This chemical is widely distributed throughout nature. In animals it is used to manufacture DNA, and any excess is converted into uric acid and is excreted. But in plants, xanthine acts a little like a convenient table on which to stack other things. One of the things plants stack on the xanthine table is a molecule called a methyl group.

If one methyl group is stacked on an xanthine table, it is called - drum-roll please - methyl xanthine. But the table will support more methyl groups than one. So if the plant attaches a second group, the chemical is then called dimethyl xanthine. (These are not as creative as some baby names, but certainly descriptive.) Quick! Can you guess what we call a molecule with three methyl groups on the xanthine? Very good! Certain kinds of trimethylxanthine are also known as caffeine. (There are actually several variations of this molecule that have similar effects. For simplicity, and because I am in a hurry, I will just leave it at three.)

These additions make caffeine kind of a lumpy, funny-shaped molecule. This is important because the molecule must attach itself to another molecule in the cell membrane in order to alter cell function. The two have to fit together like a lock and key. This peculiar shape, then, makes caffeine highly selective and specific as to which cells in the human body are affected. It has very little effect on most cells of the body. But to the cells that have the matching molecule, it binds very tightly.

Excessive, but normal, neuronal stimulation while awake produces yet another molecule called adenosine. This binds to other neurons and slows neuronal firing to keep neuron activity within safe limits. Caffeine mimics adenosine. When caffeine slips into adenosine's place, it acts like a piece of wood under your brake pedal in that you can't slow down the neuron activity. It doesn't stimulate at all. It just blocks the brake. Your own natural neurotransmitters do the stimulating, and driving without brakes does have its hazards.

The group of cells that are most sensitive to the adenosine brake are found in a small area of the brainstem. But those cells fan out and connect to every other portion of the brain. That is why caffeine has a particularly broad, if difficult to predict, effect. It can cause the heart to beat more rapidly, constricts some blood vessels, relaxes others, relaxes airways, and causes some types of muscle cells to contract more rapidly.

Quick now! Who was the composer who passionately loved coffee, and whose frenetic fugues most clearly capture the essence of the caffeinated experience? Bach!

FEBRUARY

Themes:
- Valentines
- Love
- Attraction
- Relationships
- Did I Mention Valentines?

What else is a scientist going to write about in February? What is more scientific than emotions, relationships, and love? Well, just about everything, I guess! But I think it's time we made a scientific examination of one of man-kind's greatest mysteries. Which is, why did anyone ever consent to marrying me?

I know, with all the celebration over Martin Luther King Day, we have sort of forgotten that two of the greatest Presidents of the United States were born in this month. February has some of the most significant holidays, but it's also the shortest month of the year.

So many events posed a particular challenge to my idea of having only five essays for each month. I solved this by writing three essays about love and one each for our Presidents. I hope I am not considered a bigot for leaving Martin Luther King out, but I couldn't find anything scientific about him. After all, he was a Reverend, not a mad scientist.

FEBRUARY

Week 1

THE RISK IN GETTING ONES WIRES CROSSED

Imagine you have built a small robotic car that is powered by two electric motors, one to each rear wheel. If the right motor revolves more rapidly than the left motor, the car will veer to the left. If the left motor is faster than the right, the car will turn right.

Imagine this car has two light sensors on the front of the car, set several inches apart. These light sensors are connected to the motors of the car and control the power to the electric motors. The more light that hits the sensor, the faster the motor turns. The right sensor is connected to the right motor, and the left sensor is connected to the left motor.

Imagine we have placed this car in a darkened gymnasium. It will not move because there is no light. Then imagine we have placed a remote controlled light bulb in the center of the floor. When we turn the light on, the car will begin to move. However, because of the distance between the two sensors, the amount of light striking the right sensor will be greater than the amount of light striking the left sensor. This will cause the right motor to revolve faster, and the car will veer away from the light until it goes exactly away from the light so that the amount of light to each sensor is equal. It will also go as far away from the light as possible until the sensors are no longer stimulated.

Imagine you are observing this with a friend from the high rafters of the gym. Your friend might say something like, "Wow, that thing really doesn't like the light! It runs and hides. How did you make it do that?" Of course, it doesn't "like" or "dislike" anything. It's a robot. It just appears to be a little like a cockroach. Stay with me here. This is actually very applicable to you.

Imagine you make one small change in your robot; you connect the right sensor to the left motor and the left sensor to the right motor. Then you turn the light off, reposition your robot in the gym, and you resume your perch in the rafters.

Imagine that when you turn on the light the robot moves, but this time it turns toward the light because the sensor on one side drives the motor on the opposite side. Your friend says, "Oh look, it likes the light and is moving towards it." But wait, something is drastically wrong. As the robot gets closer and closer to the light, each sensor gets more light, and this makes each motor go faster. The robot races directly toward the light with increasing speed. Your friend screams, "Look out! It's attacking!" as the robot hurtles into the light. Light and robot are demolished in one grand violent act. "Wow!" Your friend observes after a stunned silence. "That robot really hates the light."

Imagine you painstakingly reassemble your robot. This time you make one more tiny change. You place a governor on the light sensors so that the robot increases speed until a certain light intensity is reached. Above that intensity, the robot turns off the motor it is wired to.

Imagine that, meanwhile, back in the gym, this time the robot turns towards the light and rushes towards it as before. But as it gets close, it slows, stops, and sits staring adoringly at the light bulb - never moving a motor. Your friend observes, "Oh look, it's in love with the light."

Now, what has this got to do with you and me? Maybe nothing. But cockroaches, and most insects, have brains that connect to the same sides of their bodies. Muscles take the place of motors. You and I have crossed nervous systems. The left brain controls the right side of the body and the other way around. Does that partly explain human aggression? And is the difference between love and violence a simple breakdown of the speed governor, the braking system?

I don't know. However, if you think these ideas are intriguing, you would enjoy reading the book "Vehicles: experiments in synthetic psychology" by Valentino Braitenberg, available from MIT Press and on Amazon. But please finish this one first!

FEBRUARY

Week 2

GEEK ROMANCE

Have you heard about the two, red-blood cells that loved in vein? Well, OK, it's a dumb joke. But don't let me catch you repeating it then! The truth is I don't understand anything about the science of love. I think there are some subjects which maybe don't lend themselves to scientific explanations. We scientists need to recognize our limitations.

For example, I have never understood what it is that women find attractive in men. Honestly, what could any woman possibly see in a handsome, smiling, muscled, tanned, kid wearing tight jeans and driving a brand new 4X4 jacked-up pickup? He's probably not older than twenty five, immature and doesn't have a brain in his head. Girls, you must know that any guy with a great tan doesn't have a job!

Then, what's with the sinister dudes? Why are women so surprised when they turn out to be, well, sinister? And don't even get me started on bass players in rock and roll bands. . . .

If you girls really want true romance, you have overlooked a quiet group of substance, stability, and culture. Well, OK, it's a unique form of culture, but it is one. I'm talking about science geeks, of course! There are so many advantages to dating science geeks that it just surpasses my understanding why women aren't more enamored with us.

In the first place, they are generally available. Lacking in social skills, motorcycles, and tight jeans, they have been overlooked for so long that there is an over-abundance of them on today's market. Not only are geeks available, but other women seldom try to steal them. I can tell you from experience that my wife and I have been married for almost 50 years, and no woman has even made a pass at me. The only plausible explanation for this, of course, is that I am a science geek, and my wife has never allowed me to buy a motorcycle.

Geeks have other things going for them as well. One is that parents almost always love them. They appear harmless, often make good money, and can fix things. That is no reason to marry someone, obviously. But it does remove some of the difficulties from life while you look over the passing parade of bass players.

Science geeks are surprisingly sensitive and romantic people, once you get past their initial, social awkwardness. For example, you wouldn't want to miss out on Valentine endearments such as:

The Rosette Nebula is red.
The Pleides star cluster is blue.
The universe is expanding
Just like my love is for you!

It can be difficult to meet science geeks in the first place. They often have peculiar tastes in alternative music, so you seldom see them at concerts. They're even more rarely found in sports bars. They generally hang out in laboratories which have restrictive access, and they tend to socialize in groups where they discuss obscure and unintelligible topics. When seeking them out, you can be at a distinct disadvantage.

But here's a tip. Guys wear t-shirts with logos of their favorite bands and sports teams, thus showing that they are sinister dudes or are manly athletes, right? Well, science geeks tend to wear t-shirts with logos of software programs and science symbols emblazoned on them to show that they

are, uh, well, geeks. Since there is a convivial rivalry about these things, you could try wearing one yourself. Try something like a nice tight T with slogans like:
- I wear this shirt *periodically*
- Never trust an atom – they make up everything
- Code like a girl
- *Cole's Law* = thinly sliced cabbage

See if your tee strikes up any conversations!

Of course, the best way to meet science geeks is on the internet. Surfing the net allows science geeks to combine an activity with which they are comfortable – computing – with an activity they are uncomfortable with – socializing. Another strategy is to hang out in the junk food aisle of the grocery store.

Most importantly, though, science geeks thrive on mystery. So just keep being female, and they will be helplessly fascinated – with an emphasis on the "helpless" part.

FEBRUARY

Week 3

FUNDAMENTAL, INEXPLICABLE, ATTRACTIVE FORCES

Gravity got you down? (Sorry, Nerd joke.) But whether it does or not, there's not much you can do about it. Gravity is just there. It makes no sense to get mad about it. It's just one of those fundamental, inexplicable, attractive forces.

When humans don't know what something is, we give it a name, so we can talk about it. That's what happened when I was born. It worked so well that my wife and I did the same thing. We gave all of our children different names too. Then we spent most of the rest of our lives talking about them. There are a lot of things we give names to that we don't actually understand, by the way.

Like, why don't I fly off a planet that is rotating at about 700-900 miles per hour? Logic says we should fly off, but we don't. So man invented a force that keeps us on and named the force gravity. We can sort of understand parts of gravity. We still don't know, though, where it comes from.

By definition, gravity is the phenomenon by which physical bodies are attracted to each other. Gravity gives us weight and mass, which are not necessarily the same thing. Weight is the measurement of the pull of gravity on an object. We know that weight can change with location. Our weight on earth is not the same as our weight on the moon.

The mass of an object, however, is a measurement of the amount of matter something contains. If it contains more atoms it has greater mass, but its weight can still change with location. Heavy people can change their weight by going to the moon, but dense people cannot change their density. You know what I mean?! Anyway, gravity is one of the fundamental, inexplicable attractive forces of nature.

Magnetism was discovered by Sherlock Ohm. (Are we paying attention?) There are numerous theories, based on opposite poles and attractions, describing how it works. Yet no one has a clue as to why there are opposite charges in the first place. Electromagnetism, which is what we named this force, is the force that causes the interaction between electrically charged particles. It explains most of the common phenomenon of everyday life. The particles that make up our world often have one of two opposite charges. Opposite poles attract each other, and like charges repel.

These opposing elements are called charges in electricity, and poles in magnetism. An electrical current that runs in one direction creates a circular magnetic field around the wire perpendicular to the wire. In this way, magnets can be converted into electricity and electricity to magnetism. It's another one of those fundamental, inexplicable attractive forces.

The attraction between men and women, while not generally recognized by the physics community as a "fundamental, inexplicable, attractive-force", meets all the above criteria. It has been much studied, there are innumerable theories, but the fundamental causes of this attraction are inexplicable. Our understanding is far less precise, and prediction is nearly impossible. Control? Fagetaboutit!

The physicists don't like bringing this whole topic up. They know that, if the word gets out, that the real serious scientific questions are about sex, they will lose disciples quickly. I mean, look around. Notice who good looking women are seen with, and tell me it's explicable. (I think inexplicable in this context means "expletive deleted".)

We haven't even gotten to the real mystery. Everyone knows men are attracted to women. But, likewise, women seem to be attracted to men. Try and figure that one out. I told my daughters that men are highly overrated, and that I should know - being one. Alright, I admit they didn't listen, but you can't say I didn't warn them.

Here is the miracle, though, that shapes more men than our culture wants to recognize. It's the miracle that many young girls have never seen, and so, cannot comprehend. A man being attracted to women is one thing. That is biology and can be explained. A man being attracted to a particular woman is something else. That is something very different, perhaps almost holy. That one there, in the sweater, smiling shyly. . . . I must have that one. Happy Valentine's Day, Honey!

FEBRUARY

Week 4

GEORGE WASHINGTON IN POOR HUMOR

Writer E. B. White once wrote, "Analyzing humor is like dissecting a frog. Few people are interested and the frog dies of it." Now I have always thought that dissecting a frog was pretty interesting. But his reference to dissecting, in conjunction with humor, is not all that farfetched.

Early Greek, Roman, and Islamic physicians all thought that humans were composed of four bodily humors, and that an illness was evidence that these humors were out of balance. Remedies to illnesses were directed at somehow balancing these humors. We find their ideas humorous today, but originally humor was considered a liquid, not a joke.

Later the humor idea was simplified to mean that an ill person had "bad humor". (This seems a little more understandable. When I am ill, I am seldom in good humor.) But for several centuries, physicians thought they could make someone well by removing the bad humor. That is how the practice of blood-letting became popular.

Physicians removed one's blood in several different ways. Sometimes they used leeches, but more often they simply nicked a convenient vein. Of course, not every community had a physician, and so the practice of nicking veins often fell to the person with the sharpest instruments, the local barber. Only the barber and the butcher had a motive for buying only the best steel, and keeping it sharp.

Stopping the bleeding, after a bloodletting, was sometimes a challenge. Of course, this was before the days of hypodermic syringes and plastic band aids. So physicians and barbers often had to keep a large supply of linens on hand to stop the flow of blood. These linens were expensive, so they were consistently reused. In those days, the barber shops could often be identified by the bloodied white linens drying on the front porch. These were sometimes wrapped around the porch poles by the breezes. Perhaps you can visualize where the barber's pole image of red and white stripes came from.

The average person has about 70 ml of blood per kg of body weight. Therefore, a man six feet three inches tall and weighing about 230 pounds (104 kg) would have about seven liters of blood. Losing a portion of this amount isn't detrimental. People routinely give about 500 ml of blood, about a pint, when they donate to their blood bank. This averages out to be about a tenth of their blood volume.

On December 14, of 1799, our first president, George Washington, a man of the above proportions, rode his horse out to inspect his plantation from about 10:00 AM until 3:00 PM on a cold, wet, windy day. The following day he had a sore throat but went riding again to mark some trees for cutting. That evening he was hoarse, but in good humor (pardon the pun). However, somewhere between 2:00 and 3:00 AM on the 16th, he awoke in some distress. His humor had turned bad, and he told his wife he was not well.

When morning came he asked to have a bloodletting, a practice he believed in, from his estate overseer. Because he was in such distress, three different physicians were summoned to attend him. Each of them also performed bloodletting to remove the bad humor. Here is a calculation of how much blood was let over the course of that day.

12-14 ounces in the early morning by the overseer.
20 ounces when the first doctor arrived.
20 more ounces an hour or so later.
40 ounces after lunch.
32 ounces when a second doctor arrived about 3:00 PM.

The President was a sick man of 69 years of age, and he'd had over 120 ounces of blood, or about 3.75 liters, removed in fewer than twelve hours. That was about half of the estimated blood volume for a man his size. He died peacefully at about 11:30 PM that night, probably of severe hypotension and shock, after just two short days of illness.

FEBRUARY

Week 5
(I know February never has five weeks, but there is just so much going on in this month.)

THE ONLY PRESIDENTIAL PATENT

Who is the only president to hold a patent?

For some men, events in history conspire to reduce them to single dimensions. For example, I am defined by just a dot in history, which is about as one dimensional as you can get. Other men have been defined by war, or through their choice of work. Alvin York was a war hero. Ernest Shackleton was an adventurer.

This single dimension phenomenon has given definition to every president of the United States. The office is so powerful, and responsibilities so great, that whatever happens during their course of office becomes our perception of them for the ages. No matter how multifaceted they may have been, and what their interests, talents and abilities may have been, they are known for the events that circumscribed their office.

Patent 6469 was awarded to Abraham Lincoln on May 22, 1849. It was entitled "Buoying Vessels over Shoals." Lincoln envisioned a system of waterproof, fabric bladders that could be inflated, when necessary, to help ease a stuck ship over obstacles. He even designed a scale model of a ship outfitted with the device.

I suppose his invention makes him the "Father of Political Science", although I have never heard him credited with that. Actually, I have never heard of anyone credited with being the father of political science. I understand that several people have been nominated, but all declined. "If nominated, I will not run. If elected, I will not serve." Modern political scientists solve the problem by using either Plato or

Aristotle as their founder. Both are, conveniently, unable to protest.

Politicians are famous for inventing things. They just don't usually get patents on them. It seems, for the most part, that they humbly decline any responsibility for their inventions. So you have to hand it to Lincoln for standing up and taking credit for his idea.

Come to think of it, maybe we need a Political Science Patent Office (PSPO). Whereas a regular patent helps a person protect ideas so others cannot use them for free, the PSPO would prove ownership of ideas so we would know who to hold responsible for them.

But, I digress. Unfortunately, it doesn't seem that Lincoln's invention was ever actually built and used. This is peculiar because, as President, he could have just ordered that all ships be outfitted with his contraption and made a killing. I suppose political scientists hadn't invented insider trading yet.

I'm not sure what political science really is. Science is a process used for arriving at the truth about the real world. So is political science a method for arriving at truth about the world of politics? I can't quite put my finger on it, but somehow that seems contradictory. . . .

However, I hereby nominate Abraham Lincoln as the Father of Political Science. Perhaps no other president used his political office to advance science as did President Lincoln. On May 15, 1862, President Lincoln signed into law a bill creating the Department of Agriculture. On May 20, 1862, Lincoln signed into law the Homestead Act giving to any head of a family or to anyone 21 years of age, one-fourth square mile of free land for farming. On July 2, 1862, Lincoln signed into law, The Land Grant College Act, which established the land grant University system across the United States. Finally, the National Academy of Sciences (NAS), was signed into being by President Abraham Lincoln on March 3, 1863.

In the 1860's, agriculture was the science of the day. Lincoln's encouragement and support of agriculture is relatively unrecognized in 2012, his single-dimension phenomenon identity having been taken over by other events. In retrospect, the application of government resources to the support of science certainly qualifies as political science. Now, if we could just get science to financially support the government, instead of the other way around, a whole new academic discipline could be born.

MARCH

Themes:
- Science Fair
- Spring
- Decentralized Systems
- Husbandry
- Growth

In like a lion and out like a lamb. Except I don't write much about lions or lambs. They are way too ordinary. Only five percent of all animals are vertebrates. The rest are invertebrates, and these are far more interesting, if less visible.

For years, March always meant extra work as a Science Fair Judge. I have a special affinity for Science Fair kids. Many are there by compulsion, but those who choose to be there are pretty impressive.

The native bees are coming out in March. Most people think there are only honey bees and are surprised to find out that there are other kinds. I have been raising Blue Orchard bees for some time now. They are fascinating.

Decentralized systems are never more in display than in the spring, when dead things seem to come alive. Invisible things become visible, simple things become complex, and new life grows according to its own plan.

How did life come to be? Actually, I don't think evolution is the central, major theme of biology. That seeming heresy alone is enough to make this book worth the price. I suppose I really ought to write about lions and lambs sometime.

MARCH

Week 1

A VERY REASONABLE CHAPTER ABOUT EVOLUTION

I told my wife, "I think I'll try to write a very reasonable chapter about evolution." She replied, "Are you qualified to do that?" What? Does she think I'm not reasonable, or that I am not highly evolved?

Do you know who the first recorded geneticist was? It was Jacob in the Old Testament. His story in Genesis dates back somewhere between four and six thousand years. They didn't have copyright then, so the exact date is unclear. But in the thirtieth chapter of Genesis, it explains how Jacob made a deal with his father-in-law to work for a couple of daughters. Then there were all the spotted and speckled cattle, sheep and goats for a bonus.

The deal didn't sound like such a good one to me. I have trouble taking care of one wife and a few bees. Jacob had a secret plan though. He built corrals and controlled the breeding of the animals until the speckled ones were more numerous and stronger. Was his Father-in-law angry when Jacob left town with his daughters, or the herd?

Anyway, it's obvious that Jacob knew something about inheritance and selective breeding a very long time ago, even though Darwin gets all the credit. Darwin's even more popular than Mendel, and Mendel is the one who actually figured out the mechanics of inheritance. I guess Darwin has a better PAC than Mendel.

Everyone gets all excited about evolution, but hardly anyone agrees about what it is. Almost everyone thinks they have evolved into something better than they used to be, but no one thinks they used to be monkeys.

Some people think that evolution proves there is no God, but it doesn't say in Genesis that God was through creating life, only that he took a break after six days. Hard working people ought to get a vacation every once in a while. He probably went to Martha's Vineyard.

It doesn't make a lot of difference to me whether Jacob or nature chooses which animals get to mate. Some animals still get to have offspring, and some don't. Of course, in Jacob's case, the same person made all the decisions controlling breeding.

In nature, the offspring change the environment. If you put new animals into an environment, you get a new environment. Since it's the environment that selects survivors, it's like putting a new person in charge of breeding every generation. So the environment selects the survivors, and the survivors change the environment. It's like a dog chasing its tail.

This is one of the reasons why evolution has a public relations problem. A physical law usually allows us to predict or control events. The laws of physics and chemistry are often applicable in some very specific and useful ways. For example, the various gas laws can help us predict explosions and control machinery.

Evolution, great at explaining backwards, only predicts in a very general way. And while we can use the concepts for breeding more precisely, even then Kentucky Derby winners don't always produce Kentucky Derby contenders. Predicting the environment, or the individual organisms evolution will create in the future, is futile.

On the other hand, science is based upon reason. It seems reasonable that if I can breed an animal selectively, then the environment might certainly influence breeding as well. Of course, reason is only as good as the initial assumptions. The problem is that almost everyone thinks their present assumptions are accurate, even though those assumptions are different than the assumptions they held previously, or will hold tomorrow.

So if my wife's initial assumption that I was a good risk for a mate was wrong, then it is obvious I have evolved into the reasonable person I am today. If, however, she was right, then I have always been a highly-evolved, reasonable person. What is she thinking? Of course I'm qualified!

Have you ever read a more reasonable chapter on evolution?

MARCH

Week 2

SCIENCE FAIR

They say diamonds are a girl's best friend. I gave my wife a beautiful beetle once. It was from Brazil, and I had embedded it in plastic. I think she would have preferred a diamond. She apparently didn't know, or maybe she didn't care, that beetles and diamonds have something in common. However, diamonds have nothing in common with plastic. Maybe that was my problem.

But ask Lauren Richey of Springville, Utah about beetles and diamonds. For her, beetles have turned into gold. Or am I mixing my metaphors here? It all started with that annual spring rite called the Science Fair. Richey started doing science fairs in junior high. At first she was not terribly successful, but then she began to win regional and national awards for research projects that usually involved light and photonics.

As a senior in High School, Richey read a paper suggesting that iridescent butterflies might contain photonic crystals. She admits that she read the article mostly because of the beautiful blue butterfly on the front cover. But that stirred her interest. She approached John Gardner, a professor in the BYU Physics Department, with her idea to examine a beetle, *Lamprocyphus augustus*, a shimmery green beetle, with an electron microscope. Sure enough, the beetle exoskeleton contained structure similar to photonic crystals.

Excuse me, but when I first heard this story, I didn't know what a photonic crystal was. Apparently these structures, which are relatively rare, resemble the arrangement of carbon atoms in a diamond crystal. Their crystalline shape can affect the propagation of electromagnetic waves in the same way that a semiconductor can in a computer. This knowledge would allow a computer to operate on light waves instead of electricity, a coveted goal in computer science. While the crystals in a beetle are too fragile for use in a computer, they might serve as a template for the manufacture of sturdier structures.

But the significance of this story isn't really about photonic crystals. It is about a student who did college-level research while a senior in High School and published a scientific paper as a freshman in college. Since then Lauren, a sophomore in physics at BYU, has published two more research articles in professional journals and taken over 13,000 electron micrographs of beetles. She is presently examining beetles that contain photonic crystals of an opal-like nature rather than crystal shaped diamonds.

The International Science and Engineering Fair (ISEF), presently sponsored by Intel, is the world's largest international pre-college science competition. Individual schools run the series of local competitions across the United States and the world. Middle and high school students do research and compete for awards and prizes.

The science fair may be one of the best kept secrets among middle school and high school students. Students can receive many different awards, in numerous categories, as well as get a taste of scientific research. Many students receive cash prizes or scholarships. In some cases, like Lauren's, it can set the stage for an entire career. Parent support, interested teachers, and mentors are essential parts of this activity. It is fascinating to see what young minds can do.

Anyway, I was sure to show my wife this story since I think she might not have properly appreciated the beetle I gave her. She's probably sorry now that she knows all about photonic crystals. She didn't wear the mosquito ear rings or the silver cockroach pendant I gave her either. Wait. Silver cockroaches! That gives me an idea.

MARCH

Week 3

IN MOURNING

It has been a mild winter and spring. These sunny, spring days remind me of a time when I often went to eat my lunch at the Orchard Mesa Cemetery. I know that sounds peculiar. I was working outdoors at the time, and sometimes I was close to the cemetery at noon. One of my grandfathers is buried there, so it was a peaceful place where I could sit under a tree close by and eat lunch. The combination of beautiful spring days and my grandfather's gravesite made the experience sort of pleasantly melancholy.

I was reminded of these experiences when I discovered that one of my hives had died. I was standing in the midst of the beautiful sunlight, on a cloudless day, with a breeze blowing gently by. It was quiet, like a cemetery. Too quiet. Even early in the spring, on a warm day, there should have been bees buzzing around. The beehive has its own rhythm. While the number of bees declines in the winter and early spring, there should still be enough bees to tend the new babies that begin to arrive as early as February.

In the fall, bees reduce their numbers. They kick the males out (which is probably a different chapter), and the queen slowly stops laying eggs for the winter. Any given bee seldom lives much longer than about sixty days, so by February the number of bees in the hive is starting to dwindle.

Over winter, bees form a ball in the center of their hives and store up honey and pollen in a kind of rainbow arrangement around the center of the comb. Then, through the winter months, the bees on the outer edge are able to make short forays to the stored foods. From there the honey is passed, mouth to mouth, to the other bees. Then bees use their energy to shiver as a mass, creating enough heat to stay warm.

Being cold-blooded animals, bees cannot venture far from this ball in search of food when it is cold. As they exhaust the food at one level of the hive, the mass of bees moves upwards in unison, through the hive, to new storage areas. However, this instinctive behavior can lead to heartbreaking situations.

The queen can start laying eggs as early as February. The weather isn't very hospitable for flying that early in the year, and there are no flowering plants in bloom. So the bees have to depend on stored honey and pollen to survive. March becomes a critical period for them. With the new baby bees being born, food demands sky rocket, and the stored food is at its lowest supply. They can run out of food very easily with the increased number of mouths to feed.

In my hive, the bees had gotten off-center. Instead of moving straight up the hive, they moved over to one side. So when they reached the top of the hive, but far to one side, and there was no more food, they starved. Sadly, there were four frames of honey just inches away on the other side of the hive and more in the boxes below. They just couldn't figure it out, and it was too cold for exploring.

You can identify a hive of bees that have starved to death when you take the hive apart. All the bees will still be there, collected together in one place, usually at the top of the hive. Many of the bees will be headfirst into their cells, having died while trying to lick up the last bit of honey. It's a sign of the desperation that must have ensued in those final hours. It's a tremendously sad sight.

Sometimes life is bittersweet. Like, when on a beautiful spring day, you visit the grave of a deceased grandfather, or you find you have lost a hive of bees that you have tended carefully over the year.

MARCH

Week 4

DECENTRALIZED SYSTEMS

I watched the bird flock sweep across the sky. Suddenly they veered left in unison, then as suddenly darted to the right before swooping downward toward the ground only to rise upward and circle back again. Their movements were perfectly coordinated and graceful. There were no collisions, and the flock as a whole seemed more graceful than any single bird. They were even better than a well-choreographed dance troop.

The only thing unusual about this sight was that it was all taking place on a computer screen, and the birds looked more like little, grey, paper airplanes than birds. This is because I was watching a computer animation, not a real flock of birds. Yet I could see no difference in their behavior from the real thing.

Most of us assume that birds play a game of "follow the leader". We think the bird in front leads, and the others follow. But, apparently, this isn't so. The computer simulation was established by a set of simple rules instructing each "bird" in the animated flock to hold its position steady against all birds around it. That is, the "bird" on its left should be kept at a fixed distance and a specific angle. If another "bird" shows up on the right, the distance and angle should be maintained as well.

There is no "leader bird". Each bird in the flock follows simply by reacting to the movements of other birds nearby. Orderly flock interactions arise from these local bird to bird interactions. Much of human society seems to develop in the same way. Neighborhoods seem to function for the common good in many cases, especially during emergencies, long before large organizations can have much effect.

This is an example of what scientists call self-organization. Self-organization occurs when groups of autonomous particles interact in such a way as to give rise to organized patterns or behaviors. Birds are not the only animals to self-organize. Ant colonies, termites, beehives and slime molds all self-organize. The immune system operates without centralized direction. The development of the human embryo occurs without central control.

Central to the idea of self-organization is the concept of decentralization. Many social situations appear to occur in a decentralized manner. Traffic jams appear to develop spontaneously. Market economies develop complex behaviors and patterns that arise without leadership. Adam Smith argued against centralized control of economies more than two hundred years ago. One of the unique contributions of American political thought was the idea of states' rights and decentralized national government.

Decentralization appears to occur at certain scales. When things get too large, it becomes increasingly difficult to manage all the details. Efforts to do so often create, paradoxically, disorganization. For example, a beehive may house over one hundred thousand individuals. The queen does not tell the hive what to do. But upon appropriate cues, the hive grows a new queen and splits in two. It's interesting that the Soviet Union and IBM decentralized their management within one day of each other in the late 1980's.

The common assumption is that when something seems complex, it must have a complex explanation. However, time after time that doesn't appear to be true. Sixty years ago people thought that the gene must be a protein because proteins were the largest, most-complex molecules. We assumed that only the most complex molecules could account for the amazing diversity of the living world.

Instead we have discovered that inheritance and diversity are explained by a chemical code made up of just four elements. These elements behave much like a nested binary code with two elements dictating a small set of choices. The other two possibilities determine the final message. There is no boss in the cell!

Time after time, complex things are shown to be constructed of simple things utilized in unique ways. Organization can occur, indeed does occur, without central direction. It's almost as if decentralization is a part of the plan.

MARCH

Week 5
(If more crazy weather is needed.)

HUSBANDRY

So I was sitting on this beach in Kona and asking myself, "Why do waves wave?"

Sometimes, when you're sitting on a beach all day, you really just can't sleep any more. Then the strangest thoughts enter your head. For example, right after wondering how waves wave, I thought, "What am I doing here?"

My wife and I had never been on a vacation where we unpacked our suitcases before this trip. There isn't room in the grandkids' bedrooms when we visit family. Nor have we ever stayed anywhere for two weeks, unless you count my military deployment to Germany as a vacation. I certainly didn't.

The only beach I am really familiar with is that of Lake Bonneville, upon whose sands one can still bask, out at Cisco, Utah. The two experiences seem different somehow.

I was the only one in our group who noticed the sign, along the walkway to the beach, about spear-fishing regulations. It told which fish were legal to take and when. I didn't get to spear fish, though we did snorkel.

I noticed an over-all effort in Hawaii, perhaps more on the big island than on the other outlying islands, to retain and encourage a proper stewardship for the land and sea resources. Perhaps because Hawaiians are on islands, they are more aware of the need to preserve and husband resources.

"Husbandry" is an old fashioned word. In fact, it seems that the word "husband" is out of fashion in general. In its original sense, husbandry refers to a man who has accepted domestic responsibilities. It is related to managing affairs for the good of home and family. To "husband" is to use with care, to nurture, or to make last. To husband is to tend and care for resources, so that the family thrives and prospers over a lifetime. The Hawaiians seem to be trying to husband their resources.

What has this got to do with science? A century ago, biology was concerned with animal and plant husbandry, or the natural world. Gregor Mendel discovered the laws of inheritance while tending peas in the monastery garden. He was also a beekeeper. Darwin was a naturalist, concerned with the variety and abundance of life in the natural world. He raised pigeons in his backyard. The health sciences hardly existed then.

This all changed when the study of living things became the "biological sciences". When a field of study becomes a science, it is thought to have increased in precision, complexity and prestige. Thus the lowly becomes exalted, and the rustic becomes urbane. Like those are improvements . . . nurture becomes management. It might be compared to the difference between colleges and universities, or monuments and parks.

However, in this case, a rose by another name is not necessarily still a rose. There is a distinct difference between animal science and animal husbandry. The animal scientist sees the animal as a mini-factory of product, something to be managed for efficiency: the greatest gain, within the shortest time, with the least effort.

Alternatively, husbandry recognizes that the animal plays a role in the production of its own food as well as providing food for others. It can be used to either harvest or enrich the land. It can contribute work and effort, but an animal has benefit to the earth besides being a product. Scientists and husbands may be interested in identical systems, but may reach diametrically opposite conclusions about their fields. Science cannot replace husbandry, although husbands may use science.

    Until recently, husbandry preceded fatherhood. Now science has made fatherhood optional which, in turn, has made husbandry seem superfluous. I sat on the beach and wondered if there was still a place for husbandry. Good husbands will be good fathers. They will create new husbands and fathers, and like waves, both will move through time and space.

    Oh, good grief! I forgot to tell you about the waves. March just seems like a time when people ought to be thinking about husbandry during the new growing year. Oh, well, maybe later.

APRIL

Themes:
- Green
- Growth
- Living Together
- Counting New Things
- Mediocrity

Green is the color of April. The world is turning green and it's St. Patrick's Day, which means absolutely nothing to most Americans, but we act silly anyway. Celebrating a Saint from a distant Island with imaginary naughty little people (Leprechauns) leads to things like counting stuff that isn't there. I threw in mediocrity because I didn't know where else to put it. The whole month is a little foolish anyway.

APRIL

Week 1

APRIL FOOLS GREEN

Did you know that there is a tiny screen on your kitchen faucet? When I first discovered this, it freaked me out! Had I been drinking big chunks of something I didn't know about? It turns out that these filters serve two functions. They reduce the flow of water to avoid splashing in the sink under high pressure, but they also aerate the water. Water with high oxygen content tastes better than "flat" water. Most of these screens can be removed simply by unscrewing them. That brings up an interesting possibility.

There is a cute, little product on the market called "Tub Tint". Tub Tints are commercial versions of "fizzies" that are made of various combinations of baking soda, citric acid, corn starch and salt. They fizz in water and turn the water different colors. You can make your own versions pretty easily. Do a web search on "fizzies". However, homemade versions are usually too large for the following scientific experiment. You'll need tiny tints for this.

Anyway, if one were to unscrew the filter and slip a green Tub Tint into the faucet and then replace the filter, the water would run green. I am not sure if Tub Tint comes in green. But you could always put in a blue and yellow for the same for the same effect. Of course you could do other colors too, but green seems the most appropriate for April. Do not use Easter egg dye or food coloring unless you want stained skin. No one would want to do that, would they? In fact, I can't imagine why anyone would want to do this whole experiment. Idle curiosity maybe.

Another interesting and relevant chemical phenomenon is that the copper compounds in a variety of algaecides used in aquarium treatments will turn blonde hair green . . . if it somehow got into someone's shampoo. Hey, I'm not suggesting anything. I'm just passing on some interesting, scientific facts.

    Did you know that alcohol mixes with lipids and fats? That's one of the reasons it is absorbed so quickly into the body. Cell membranes are partly made of lipids, and ethanol moves directly through the membranes. There are heavier alcohols, like methanol and propanol, that are so soluble in lipids they don't just move through the membrane. They dissolve the whole membrane! This is not healthy. Do not drink these.

    But heavy alcohols can have their uses. Because they dissolve in lipids, they are good solvents for cleaning up after oil and fat stains. They also burn pretty well which is why they are sometimes put into gasoline additives like "HEET". I guess that is supposed to help the gasoline burn and clean all the fat out of your engine. I didn't even know engines had fat, but that does bring up an interesting possibility.

    Pure chemicals generally burn with characteristic colors, so you could use methanol (HEET) to set fire to something like Boron. All you need is a little Borax, which contains Boron, and which you may have in the medicine cabinet or can buy in a drug store. Simply pour a little HEET into a fire-safe container and add a little Boron. A half a cup of HEET and one or two tablespoons of Boron should burn with a green flame for about ten minutes. I don't know if the green fire attracts or repels the "little people". All of this is just theoretical, you understand.

    Have you ever wondered what might happen if the ink from a fluorescent highlighter accidentally was mixed into a glass of water? Neither have I. That could sure be a mess. But if such a thing happened, the water just might fluoresce. I wonder what would happen if someone put a white flower in that same glass of water. Would the flower take up the dye and the petals fluoresce under black light? There's only one way to find out. If we knew what we were doing, it wouldn't be called an experiment.

APRIL

Week 2

THE GOOD, BAD, AND UGLY

I have good news and bad news. Which do you want first? Wait! This is my book so I get to decide. The only reason I asked was that I'm not sure which news is good and which is bad.

The words "good" and "bad" do not have strict technological definitions. In fact, the two words probably describe a spectrum in which some things are perceived as good, but something else is perceived as either better or worse. They are values, not absolute facts. Keep that in mind when you read health newsletters.

There are things that most all humans agree are good and bad. Most humans think killing other humans is bad unless it's for a cause that we think is worse. It seems most humans desire love and affection, although I sure can't figure out some people's taste. Why doesn't everyone love me?

Science, of course, doesn't deal in values. Science deals with the physical world and matters that can be verified with empirical data. In fact, some scientists deny that there is any other kind of world and insist that consideration of things like religion and morality are bad. These are often the scientists who claim that we should fund scientific research because science does so much "good".

Obviously the same science that brings us electric lights also created the electric chair. The science that discovered antibiotics has created the antibiotic-resistant bacteria that now plague us, as well as poisonous gases. The same science that created the nuclear bomb now provides us with news about nuclear catastrophes by way of electricity generated in nuclear power plants.

It isn't that science that is inherently good or bad. It's how the science is applied that can be considered good or bad. Some scientists call religion bad but fail to recognize that it is not necessarily the religious belief they are judging. As with science, the application is what may be faulty.

Did you know that generally a person with a common cold heals cuts faster? So are colds bad or good? I suppose that depends on how seriously you are cut. If it is just a scratch, then no one cares, and the cold is miserable. Miserable being another word for bad. If the carotid artery is severed, a cold is the least of your worries. But there might be an advantage to having a cold if you have a serious cut on the arm. The faster it heals, the less chance of infection. Colds activate many healing powers of the body that by chance also enhances wound healing.

The parasitic ameba, *Entamoeba histolytica,* can make you and I deathly ill, but millions of people seem to harbor it with no apparent ill effects. So is the parasite bad or good? It just isn't clear.

I much prefer being healthy. Healthy is defined as not having anything abnormal, out-of-order, or bad going on. Health exists when everything is good. However, there are numerous things in my life that are bad: work, deadlines, meetings, broccoli . . . . So am I healthy? If I am not, then I must be unhealthy, or un-good. Isn't that the same as bad? It is if I am eating broccoli.

If I have a disease, I am not at ease, and that's not good. So it must be bad. So is broccoli a disease? Of course. But because bad is a relative term, there are some things worse than broccoli. No, I'm being serious here.

Is high blood pressure worse than gout? People with high blood pressure are at risk of a heart attack and death. Someone having a gout attack may not die although they just might wish they could. Hydrochlorothiazide is a commonly prescribed water pill for lowering blood pressure, but it is one of the major causes of gout attacks. So is Hydrochlorothiazide good or bad? Sorry! Science doesn't deal in value questions.

In conclusion, we have twenty percent of the American economy devoted to health care. Is that good? Well, I have good news and bad news. Which do you want first?

APRIL

Week 3

LIVING TOGETHER

I watched my little bee break out of its cocoon today. I was surprised at how proud I felt over her success. The male hatched out while I wasn't looking, but I did see him for a few minutes. Man, are they fast! When he left I could hardly follow which direction he went.

You probably didn't know that bees have cocoons. Most folks think all bees are honey bees, the kind that live in hives and make honey. (Do honey bees "make" honey, or do they "gather" it? Technically, honey is nectar from plants, so bees simply gather it. Of course it is changed while in the bee's stomach and then stored in the hive. So I guess you could say they make it too.)

But my little pet bee that hatched out today is what is sometimes called a solitary bee because each female bee builds a single nest, deposits her off-spring with provisions, and then dies. She works alone. I like to call them native bees because they were the only bees on the American continent until the early pilgrims brought the honey bee with them across the ocean. Specifically this little bee that hatched out of its cocoon today, in mid-March, is named *Osmia lignaria* by the scientific community. Others call it the Blue Orchard Bee or the Mason Bee.

She's a cute little bee that doesn't look much like a honey bee. To begin with, she is black, or a very dark blue depending on the lighting. She is also smaller than a honey bee and grows to be only about a quarter of an inch in length. Somehow she is endearing in a way I didn't expect. I can't quite explain it, but she was just cute in a way that I had never thought about a honey bee. She sat for a few minutes, groomed some stray hairs, walked a few paces, defecated following her long winter in the cocoon, sat in the sun for a few minutes, and then, in a flash, was gone.

We had lived together all winter by then. I got my bees last fall and have been carefully storing them; first outside, then in a refrigerator all winter. My plan is to release them this spring, let them pollinate some orchards, and then collect their babies for release the next year. I guess it is sort of free-range bee ranching but without the branding and roundup.

I have been fascinated by things "living together" since my senior year in college when I took a course in Parasitology, the study of parasites. I know that sounds gross, but I found the concept of things living, cooperating, and adapting to live together especially fascinating.

In fact, every animal ever examined has at least one specific animal that lives exclusively in, or on, the host. In addition every animal examined shares some collection of animals that live in, or on, it with some other species. Inescapably then, there are more "parasites" than free living animals in the world. The scientific term for these co-dependent creatures is "symbionts", and most symbionts do not cause disease or in any way harm their hosts. Many benefit their hosts and are in turn benefitted.

Bees aren't parasites, but they are symbionts. Their entire lives are intertwined with the flowering plants that provide them both with pollen and nectar. In turn the flowering plants are entirely dependent on the bees to provide the very intimate service of reproduction. Or maybe flowers are the symbionts of the bees? Sometimes it is hard to tell.

This mutual intimate relationship between insects and flowers benefits humans with the very world in which we live.

About eighty or ninety percent of all flowering plants are pollinated by animals. There are about 200,000 animal pollinators in the world, and, of these, the great majority are insects. The most successful insect pollinators are bees. Without pollinators, the world as we know it would simply cease to exist.

So in yet another sense, I have been living with my little Blue Orchard Bee much longer than just one winter. I have been living with bees all my life. In fact, I owe my very existence, at least as I now exist, to these marvelous insects.

And that's why I have my bees. If I can create a home for these little creatures, I create a small part of the earth for me. When I tend the bees, they attend to my needs by providing sweet fruit, healthy vegetables, new seeds and beautiful flowers. Just maybe, by building a better world for myself, I also create a better garden for my neighbors, more flowers for my community, and a better world for our planet. Not a bad deal. And, on top of it all, I got to watch my little bee hatch this morning.

APRIL

Week 4

COUNTING THINGS THAT AREN'T THERE

Much of science seems to revolve around counting things. So why are accountants called ac-count-ants, and scientists are not. I think, if I had it to do it over again, I might be an accountant instead of a scientist. Counting money has accounted for much of my time anyway. And accountants seem to have more of it to count than I do.

If I had been an accountant, I would have missed out on counting so many other interesting things. I have counted nematode eggs, salvaging ants, numbers of mosquitoes, the number of eggs laid by solitary bees, and how many degrees centigrade must accumulate for a mosquito to hatch. I've even counted the number of students who finish an exam over set-timed periods. Great interesting stuff!

One thing bothers me though. Scientists sometimes count things that aren't there! How do they do that? Of course, accountants do the same thing. How do accountants account for the number of people who are un-employed? Even more peculiar, how do they know how many people have "quit" looking for a job? How do they count the money that wasn't spent during last year's Christmas season?

Did you know that counting things that aren't there is a very ancient tradition? Robert Hooke was a brilliant scientist in the 1600's. He invented air pumps, diving bells, rain gauges, and even a mechanical adding machine. He wasn't particularly well liked because he was so odd and had a nasty temper. He and Sir Isaac Newton had a running feud for years. (That's another question. Why are so many scientists such odd characters? I am about the only normal one I know.)

Oh, back to Hooke. He was examining a piece of cork with a microscope. (Let's hear of an accountant doing something interesting like that.) He could see that it was made up of tiny perforations, or holes, so he decided to count them. Here is a quote from his paper on the subject.

"I told (counted) several lines of these pores and found that there were usually about three score of these small cells placed end ways in an eighteenth part of an inch, whence I concluded that there must be near eleven-hundred of them . . . in the length of an inch, and therefore . . . about 1200 million (in a cubic inch)."

I am reminded of a line from a Beatles song "Now they know how many holes it takes to fill the Albert Hall". I mean Hooke was counting holes for Pete's sake! There's nothing there. The Albert Hall is an empty space. It's like asking how many nothings fit into nothing.

Still, the quote above is important because it was the first known use of the word "cell" to refer to the structure of living things. It was two centuries before Theodor Schwann proposed the cell theory which states that "All living things are composed of cells and cell products that are reproduced". This is more or less the root of our understanding of living systems today.

Even stranger than "counting nothing" are the physical scientists who count things that can't even be seen. I was suspicious when I first learned that chemists counted atoms. Now they count electrons and quarks. Come on! And Physicists count magnetic forces, lumens, and even calories.

Mathematicians, however, account for the strangest counting. They count numbers which are nothing more than an idea to begin with. This leads them to counting such things as groups of ideas, which are simply the idea of a group of ideas. For example, do you know how many numbers are "even prime numbers"? Just one. It's two. Oh, and there are five "platonic solids", whatever those are.

    I'm not totally clear on the advantage of knowing that the number of Archimedean solids is thirteen. But how many grains of pollen a bee can carry in its pollen sac? Now that's something you can count on.

APRIL

Week 5
(Don't worry. There is no extra charge for these extra chapters.)

ODE TO MEDIOCRITY

Personally, I have always thought that mediocrity is undervalued. This is especially true in the biological sciences. No two living things are ever exactly alike. It's almost eerie that way. I mean two hydrogen atoms are just alike, and a volt is always a volt. But two canaries aren't identical. How can anyone make sense of that?

Well, the way we make sense of variation is to look at averages. The average is, of course, the central tendency of a set of data. To get an average, we add up all the values and divide that number by the number of values. If we add all the test scores and divide by the number of tests given we get the average test score. And you can see at a glance that average is decidedly mediocre.

There is a reason for all that variation out there you know. If all things are alike, and the environment changes in a way that is harmful, then all identical things will be equally harmed. But variation helps to assure that some living things survive to make more living things. In order to talk about a category of variable things, we calculate the midpoint of their characteristics, and call the midpoint the average. But we all know that, in human talk, average means mediocre.

If all humans were just alike, it would be difficult to have a decent conversation. But when we are all different, we can talk about the other humans that aren't there. But what if we want to identify all humans by category? Well, then we must look to the mediocre average, and, as you shall see, that is no small thing.

The mediocre are the elusive average, the theoretical mean, the standard upon which all men rely for comparison. Think about it. Without mediocrity there can be no upper class to tax. There would be no tired, poor, and huddled masses to receive charity or government assistance. In fact, what would government even do without those who have below-average income?

Without the mediocre, there would be no poverty level from which to claim privilege, no superior Olympians or Rhodes scholars, no maniacs who drive faster than I do, and no idiots who drive slower than I do. Why, we couldn't even have grade inflation without the average student. To give credit where credit is due, the silent and enduring mediocre should be hailed as the new, true elite!

Morality and ethics would no longer be debatable issues. I mean, after all, most people are, by definition, just average. Mediocre is about as close to being humble as you can get without actually being it. One doesn't really have to worry about pride when you are mediocre. Being average is especially great because, since you aren't truly humble you don't have to worry about the inheritance taxes the meek are going to have to deal with. (Whoo boy, now there's a problem I wouldn't want.)

Anyway, not many people want to admit to mediocrity, so declaring averageness demonstrates ones true honesty and sensitivity. No one else really thinks they are less than average, in spite of the fact that by definition half of everyone around you is below average. So being average makes everyone around you feel good to be superior to you. The way I see it, by being mediocre, I am actually in the service of building up other people's self-esteem. Is there any higher calling than being in the service of your fellow man?

Obviously mediocrity is the focal point upon which all understanding revolves. Those who wish to be prosperous, healthy, or good looking are simply being presumptuous. Why should they be any better than the mediocre from whom they are dependent for their very identity? No, it is in that magical mode of mediocrity where normalcy lies. The world rests upon the broad shoulders of the average. The mean is the Atlas of our modern world.

MAY

Themes:
- Mother's Day
- Rising Rivers
- Testing
- Summer Hazards
- Gardens

Now this is my kind of month! All kinds of things are going on. It's time for final exams, my wife wants the air conditioner running, the rivers up, and don't forget Mother's Day. There are graduation exercises with the accompanying graduation lectures, preparations for June weddings, getting the garden started, and so much more. May might just be the busiest month of the year besides December and back to school in September.

MAY

Week 1

MOTHER'S DAY SCIENCE

Mom to two year old: "Don't touch that! It's hot."

Two year old thinks: "What's hot?"

What has Mom just done? She has almost guaranteed that the child will touch whatever it is that is hot. This will be called an experiment when the child gets older and is looking for ways to justify dumb behavior. Mom thought she was teaching, and I suppose she was. She was raising curiosity levels in the child, so the child could learn.

A child does not know what "hot" is until he or she has felt "hot". There really is no other way to do it. You can say, "It will hurt", but at some point the learner has to know what hurt feels like. It won't do much good to physically keep the child from touching something hot. Sooner or later, they have to experience "hot" in order to live wisely. The best Mom can do is keep the risk for serious harm low.

We could think back even further. A baby knows whether it is hungry or not. Its response to any kind of discomfort is to scream bloody murder. Of course, most adults are far more sophisticated than that. When adults aren't happy they just riot in the streets and break things. They also scream "bloody murder" or some radical saying, like "Occupy", "Pigs in a blanket", or "Grabblefrack".

Only after time and experience, does a baby learn that the best source of comfort of all kinds is that one special person. It is even later that the child learns to call her Mother. The best that Mom can do then is provide protection and encouragement, but refuse to give in to tantrums. Oh, and maybe a little food.

Our first experiences are often misleading. Mom says not to touch something, but she herself touches the same thing all the time. Apparently, sometimes it's hot, and sometimes it's not. Life can be so confusing. How does she do that anyway? Watching Mom work the magic of not getting burned while working a hot stove is fascinating. The child thinks, "I wonder how she does that?"

Teachers often think they are teaching when, in fact, students have no clue what the teachers are talking about. This occurs at all levels of education, even with adults. Education is acquired, not by listening to words, but by experiences in the environment. The words simply help us name the experiences and be able to manipulate them in our minds.

Yet in our public education system, from pre-school up, we continually tell people "Don't touch that" when the students are actually thinking, "What's that?" At other times we tell students, "That is good for you." The student may not have a clue as to what "that" is, or how to obtain it. Further, what is good for them seldom seems very much like it is at the time.

It's hard to think very clearly about gravity until one has fallen out of a tree. Experience with magnets is necessary somewhere along the way if one is ever going to understand electricity. Chemical reactions are easily visible, even if atoms and molecules are invisible. Teachers of science are lucky because it is not difficult to arrange for experiences with the real world.

It us much more difficult to imagine abstract concepts without physical experience. For example, economic concepts may be difficult if one has never earned an income. How does one learn to manage money without ever having lived within a budget? Well, you could always ask congress, I suppose. They must know.

You Mothers may not realize that you hold the key to America's economic well-being and educational superiority on the world stage. It's the mothers who stimulate children to learn from their physical experiences. It's my Mother telling me not to touch things that were hot that started me wondering what hot was.

My Mother also told me that "Can't never done nothing". I pondered the meaning of that often as a child. When I complained about going out in the rain, she told me, repeatedly, that I wasn't "sugar, or salt, or anybody's honey." I mean statements like that just beg for experimentation.

MAY

Week 2

RIVER RUNNING

It's May and the river is up. I have a friend who is a river runner and knows a lot about water. I'm not exactly sure what a river runner is. Personally I can't even walk on water on a still, small pond, but he can apparently run on the river. Anyway, when we went to dinner the other night he told me that there was 10,000 cubic feet of water flowing in the Colorado River that day per second. They expect there to be as much as 35,000 feet per second flowing at the river's crest. He wondered how many elephants that would be equivalent to in weight.

I don't know too much about elephants, so I looked up what a cubic foot of water was. One cubic foot of water per second is 7.48 gallons, and a gallon weighs about eight pounds. A five-gallon bucket of water would weigh forty pounds. I happen to know that a five pound bucket of honey weighs about sixty pounds, so I guess honey is thicker than water. But I have no idea how much a five pound bucket of blood would weigh, so I can't really verify that blood is thicker than water.

Anyway, eight pounds times 7.48 gallons of water would be 59.84 pounds of water in one cubic foot per second. So, 10,000 cubic feet of water passing by a given point at any given second, would weigh 598,400 lbs.

According to the Mozilla Firefox search engine, the African bush elephant weighs up to 16,500 pounds. (I think I'll round it off at just 16,000, knowing how mammologists always exaggerate.) If we divide 598,400 lbs. by 16,000 we get 37.4 standard elephants worth of water that passes a given point every second.

That's impressive! I think if you were running on the river, you would definitely want to stay on top because if you were on the bottom things could get pretty heavy. Granted, the weight would be spread out over the surface of the river bed. So if you were to go under the water, it would be better to do it where the river is wide. That would provide an increased surface area for the weight to rest on. My friend explained to me that, unfortunately, that is seldom where people go under.

There is another problem also. These calculations are just the weight of elephants on top of you at one undefined point for one second. I mean there would be 37.4 more elephants on top of you in just another second. I believe that could easily be called a stampede!

But how big is a point? (I was about to ask, "What is the point?", but decided against it.) Suppose you were to get in front of all the water coming down stream. (I really don't see how you could get behind it very well, can you?) You would then have the weight of all the water up stream pushing you downstream.

Let's suppose a point is six-feet long. That's how tall I am. You would have the first 37.4 elephants pushing down on you. Behind that there would be another 37.4 elephants pushing you downstream. Behind that would be another 37.4 elephants, and so on upstream. In just one hundred yards you would have the equivalent of 623 elephants (or 9,973,333 pounds) pushing you downstream.

I can see right away, even if my assumptions are a little off, that walking on the water isn't going to do anyone much good. If you are going to do anything, you had better run on that river, and run like hard. I guess that's why they call them river runners.

I want to thank my friend for his help in writing this

article. I couldn't have done it without him. In fact, I would never have thought of such a dumb question by myself. Do you think river runners are a little strange?

MAY

Week 3

DRUG EVALUATION TESTING

Due to lack of sleep and stress, students often become ill during finals week. In an effort to find a way of alleviating this on-going problem, I recently conducted a study during final exam week at Colorado Mesa University. We tested the use of placebos, extra strength placebos, and generic placebos as final exam disease preventatives.

The placebos tested all contained the active ingredient $C_6H_{12}O_6$. The placebos and extra strength placebos were obtained from a classified ad in The Daily Sentinel. The generic placebos were obtained from City Market from the sugar aisle under the label "Sugar Cubes". All pills were transferred to old, left-over, empty prescription bottles. The old labels were removed and no marking was left except the gross sticky stuff that never comes off of those things. Each bottle was then hand encoded with the labels "p", "esp" and "gp" respectively to disguise the contents. This code was given to a retiring colleague, to await the conclusion of the trial.

Several students were recruited to evaluate the effectiveness of the pills by promising one letter grade improvement on their Anatomy and Physiology course for participation. Seventy one people volunteered, the entire class, but only seven were eventually chosen. The rest were already sick. Those participating appeared healthy and rested at the beginning of the study, with the exception of Ashley, who appeared to be slightly hung-over.

The subjects were supplied the pills at no cost to themselves. However, to make the test more real, they were asked to imagine that they were paying five dollars per pill for the placebo, ten dollars for the extra strength placebo and one dollar for the generic placebo. However, at the end of the study three students reported that they had not actually done this. They thought it was all free under the new medical care plan.

Subjects took three pills a day, one with each meal, for one week prior to finals and throughout final week. This created some experimental design problems because I forgot that college students never eat three meals a day: sometimes it's one, sometimes seven. Some subjects took pills with all meals and some only took them when they were sober. Another confusing variable was that several students counted cokes and candy bars as meals.

One subject taking the generic placebo dropped out of the study when they discovered they were flunking no matter whether they took their finals or not. Another subject withdrew from the study claiming the placebos caused insomnia, nerves, sweaty hands, dry mouth, and headaches. I suggested she try putting her hands in her mouth and continue with the study. She suggested I could do something else.

At the end of the study, all students reported that they thought the extra strength placebos were the most effective, followed by the placebo, and then the generic placebos. In fact, they estimated that the generic placebo was only half as effective as the extra-strength placebo. However, since the generic placebos were the most pleasant to take, and because of the price differential, most thought they would just double the dosage on the generic compound to bring their potency up to the level of the other pills.

In contrast to the self-reporting data referred to above, it should be noted that all participants were still sick by the end of final week making it difficult to assess the effectiveness of administering placebos as a disease prevention strategy.

Some side effects were reported such as gastrointestinal distress and weight gain. However, the test was terminated before those claims could be substantiated. To evaluate these claims would require a much more controlled trial as it pertains to diet, drink, and other extra-curricular activities.

In conclusion, we can conclude that finals week is stressful for teachers as well as students. It is recommended that college professors not be held accountable for any written work during this time of year.

MAY

Week 4

CAUSE AND EFFECT

It was clearly my wife's fault that I sliced open my hand. If she hadn't scheduled a party for 21 little girls at our house, I wouldn't have been on the roof trying to get the air conditioner operating in such a hurry. The wet roof, the slick shoes, lack of coordination and balance, and working too fast had nothing to do with it.

The determination of cause haunts science. It is a surprisingly difficult concept to prove. Even when the accident was clearly because driver A ran a red light, there is a reason that driver A ran the red light, and a reason driver B was in the intersection. What was the real "cause"? Children quickly learn to use this slippery concept when they claim, "He hit me first", or, "I didn't turn my homework in because the dog ate it." It's never anyone's fault.

The determination of cause is an especially difficult problem with disease because the assessment of exposure to a disease agent is often imprecise, and the mechanism that connects exposure to outcome is not yet known. Consider the effort to prove that cigarettes cause some forms of cancer. It took a long time to prove even though everyone was already pretty sure that they were somehow bad for you.

Honey Bee Colony Collapse Syndrome has led to massive declines in the number of bees in recent years. This is significant because honey bees are the major commercial pollinators in the world. Without their pollination services our world would be altered radically and detrimentally. But the cause of this bee disease has not been known although there have been numerous hypotheses.

As is often the case, there may be more than one contributor to the problem. It now appears that colony collapse may be caused by an imported, emerging, bee disease, *Nosema ceranae*. Three recent papers by Dr. Mariano Higes from the Bee Pathology Laboratory in Marchamalo, Spain suggest that a new species of *Nosema* has spread throughout the world and may be causing the disease. Dr. Higes has identified *Nosema ceranae* in several commercial outbreaks in Spain where they were able to rule out numerous other potential causes such as pesticides or other diseases.

In a companion paper, he and his research team also found *N. ceranae* causes death in Bumble Bees, another important pollinator. In his third publication, they were able to show that *N. ceranae* has developed into an important pathogen because of better reproductive response at higher temperatures. So the cause of Colony Collapse Disorder may be *Nosema ceranae*.

The question, then, is if *Nosema ceranae* causes colony collapse disorder, what causes *Nosema ceranae*? That may be the fault of Dr. Robert Atkins and Dr. Arthur Agatston; the creators of the Atkins diet and the South Beach diet respectively. Both have recommended almonds as healthy snack foods. For that, and other reasons, almond production has exploded in recent years. In the United States, almonds are grown mostly in southern California.

First the expansion in almond orchards surpassed the ability of local bee keepers to pollinate the crops. Then bee keepers began trucking bees into southern California from all across the nation. Eventually some bees were brought in from New Zealand. Unfortunately, some of those bees had originated in China where *Nosema ceranae* has been in existence for a long time. Chinese bees have become more or less resistant to it. But the disease spread into susceptible American bees and was then disseminated across the nation by the migratory pollination practices of commercial bee keepers.

    Whatever caused these two physicians to recommend almonds to the public? It was partially due to concern over obesity in modern Americans. Obesity appears to be largely due to over eating. Interestingly sugar consumption is the main culprit. Sugar consumption has increased historically because it was cheaper than honey. Honey is expensive, in part, because of bee diseases. Kinda makes a person dizzy, don't it?

MAY

Week 5
(We may need a week five in May.)

MINIMAL MEASUREMENTS

I used to help my Grandfather in his garden, at least until he got tired of my help and sent me home. He had a large, wooden frame with quarter inch mesh wire on it that he kept on the edge of his garden. Every day, he threw several shovels of dirt onto this frame. The small stuff would fall through, but the larger rocks and sticks would be retained to be hauled away and discarded. This way he constantly improved the texture of the soil in his garden.

This is a pretty easy concept to grasp. Anything that could be broken up into pieces smaller than a quarter of an inch would pass through. Obviously, a smaller mesh could be used and yet smaller particles separated. The smaller the mesh the smaller the particles. It wouldn't make any difference if it was one eighth-inch mesh, or one sixteenth-inch mesh except the particles would be smaller. So, would it still work if it was one five hundred and twelfth inch mesh? Of course. The particles would just get smaller.

Humans can only communicate and work together because of shared experiences. When we use common measurements, we assume that others have had experiences with those units of measurement. If I tell you something is about six blocks away, we may not have the exact same image in our minds, but we are pretty close. If I tell the barber to take an inch off, it isn't likely he or she will take off three inches. The more we share our experiences, the easier it is for us to agree on what to do, and how much of it to do. Lacking shared experiences, we often disagree about goals as well as methods used to obtain them.

Most Americans measure in inches, feet, yards and miles. But in science we most often use the metric system. I don't think it is because of some ultimate superiority, although some will argue for one system or another. But in science we often need to communicate across language and cultural barriers. Having a shared system of measurements helps us work collectively.

So when we start talking about screens that have very tiny mesh openings, we usually use the metric system. A meter is one hundred centimeters and approximately a yard. A centimeter is ten millimeters or around a quarter of an inch. A millimeter is similar to a sixteenth of an inch, so you can see that a millimeter is getting pretty small for use in our common every-day experiences. Can you envision a millimeter or sixteenth of an inch? Now imagine that you could etch 1000 little lines, like ruler lines, in that tiny space. Those little units are called micrometers and we use the symbol um. Most cells are measured in these units. For example, red blood cells are usually about seven um.

Can you imagine etching 1000 more lines between the two lines of a micrometer? If you can, that unit is called an Angstrom. Cell membranes, internal cell structures, viruses and even some proteins are measured in these units. But most of us don't get any experiences with these measurements.

A team of engineers and chemists have now created a computer chip that works about like a coin sorter, only with much smaller openings. Liquids flow across a screen where big particles get stuck and smaller particles pass through a slot in the bottom. That slot is measured in Angstroms. Each chip has a smaller and smaller slot until a single particle can be trapped. They have dubbed their creation a *Lab-on-a-chip*. Once the particles have been sorted by size they can be examined with an electron microscope to determine their identity. Such a chip could greatly enhance early virus detection in patients. It could also be used as a research tool for virus purification. It's a great invention, but it isn't a lot different from my grandfather's garden screen.

JUNE

Themes:
- Bee Venom
- Romantic Poetry
- Insects
- Art
- Jell-O

This is the month of nectar flow. School's out, and I have to play catch-up with beekeeping. That usually means a few stings. Lots of weddings in June, and a young naturalists fancy turns to . . . insects of course.

Something about the time out of school, the growing world, and nice days sort of inspires art and Jell-O. I especially think of Knox blocks. Anyone remember this cool summer treat? It was the original inspiration for seismographs.

JUNE

Week 1

BEE VENOM

Last week I received 0.1 mg of bee venom injected into my chin. I'm pretty confident of the amount because that is how much the average venom gland on a bee stinger holds. I suppose there are some size discrepancies between bee venom glands but I'm not sure how significant the differences would be. However, I would have preferred less.

When a bee stings, it doesn't immediately inject the entire amount of venom. The bee stinger has a barb on the tip, so that once it has penetrated, it can't be easily withdrawn. Hence when the bee attempts to leave its victim, it pulls the venom sack out of the bees' abdomen, and the sack remains attached.

This venom sack has a type of muscle surrounding it that continues to pulsate and pump venom through the stinger and into the victim's skin. I am told that with a little presence of mind you can watch this phenomenon happen. I've never been present in my mind at such times. If the stinger remains attached for more than a few seconds, the entire 0.1 mg of venom will be injected. However, if you can get the stinger out of the skin very quickly, you will only receive a portion of the venom; and the effects of the sting will be lessened.

The simplest way to accomplish this is to quickly scrape the stinger out using your fingernail. In my case, the veil that was supposed to protect my face was in my way, and I had gloves on. So my fingernails were not readily available at the moment. It was some time later before I could even try to remove the stinger and venom sack. I'm pretty sure the sack was empty by then. I got the whole 0.1 mg.

There are lots of suggested antidotes out there for bee

stings. Mostly they suggest basic compounds to be applied to the outside of the skin. It's doubtful that any of them work because the venom is under the skin. However, cold compresses may be soothing and helpful by limiting how far the venom spreads. Some say alcohol alleviates symptoms, but I don't think it's wise to have alcohol in the bee yard for a variety of reasons. No, I think you are supposed to apply it to the site, not drink it.

Honey bee venom is a colorless liquid that tastes bitter. I guess you would have to milk several bees to get enough to taste though. It is a complex mixture of at least eight proteins. These include hyaluronidase which dilates capillaries so the irritation spreads out quickly. Dopamine and adrenaline are present. They say these are what increase ones heart rate after a sting, but I think this happens due to the pain alone. Man, I love this job!

About half of the proteins are a compound called melittin which is a strong anti-inflammatory agent that causes the release of cortisol. Cortisol, in turn, is usually released in response to stress, which seems appropriate here. However, it inhibits the immune system which may help explain why bee stings are irritating for several days.

The second most abundant protein is an enzyme called phospholipase. Phospholipase has several different functions. It lowers blood pressure, although I failed to notice that effect this past week. Phospholipase also destroys cell membranes in the local region and inhibits blood coagulation. This makes the effects of the sting spread out from the site of the puncture, sometimes encompassing entire chins.

Now, I apparently have a weak chin because my wife immediately noticed my augmented chin when I came in the house. She commented that I actually looked a little better that way and seemed disappointed that it was only temporary. When asked specifically, though, she denied that. She is a kind person. How else do you think she ended up married to me? Uncontrolled compassion. Later she started giggling at the dinner table and accused me of having Beetox treatments. Cute!

JUNE

Week 2

PO-ENTOMOLOGY

Recently one my student's obviously confused two different parasites in an answer on a test. The student jokingly claimed "poetic license". That made me think about the poetry of science. It was a brief thought because it is a brief subject. But it turns out that poets have often been fond of scientific metaphors, or at least insect analogies.

For example, Thomas Hood, a British poet from the early eighteen hundreds used an insect to create a sense of the supernatural in his poem, "The Haunted House". He wrote, "And on a wall, as chilly as a tomb, the death's-head moth was clinging." It should be noted though, that these moths are simple, nocturnal creatures that mostly feed on nectar. They can invade bee hives and feed unmolested because they can mimic the smell of the bees. But they never feed on dead heads.

Insects are also used to make philosophical statements. Although not exactly poetic, Solomon admonished us to, "Go to the ant, thou sluggard; consider her ways, and be wise." I think this means that lazy people should be working harder. But considering how dumb ants are maybe he means don't be stupid and work so hard. With poet types, it's sometimes hard to know what they mean.

For example, Anton Chekhov, the Russian playwright and short story master, once said, "In nature a repulsive caterpillar turns into a lovely butterfly. But with humans it is the other way around: a lovely butterfly turns into a repulsive caterpillar." I'm not sure what he means by that, or what evidence he might have been considering as to whether or not that is true.

Poets are assumed to be wise and their words loaded with meaning. However, I've noticed that they frequently make mistakes when they write about insects. It is perhaps the most disheartening to discover that the Bard himself, William Shakespeare, for which I have a lifelong appreciation, made entomological errors. In Henry V, he says of bees, "They have a King, and officers of sort." And then he compounded his error in the line, "To the tent-royal of their emperor: Who, busied, in his majesty, surveys the singing masons building roofs of gold;" In this case the head man is actually a queen, and the masons are all women.

Oliver Wendell Holmes wrote, "Thou art a female, Katydid! I know it by the trill that quivers through thy piercing note so petulant and shrill." Unfortunately, only the male Katydid sings.

On the other hand, sometimes the poets are the first to get it right. Sir Ronald Ross discovered that the mosquito was the vector for Malaria in 1902. For this he won the Nobel Prize and became quite famous. However, Henry Wadsworth Longfellow alluded to this relationship between the mosquito and malaria in his epic poem, Hiawatha, published in 1855. In this poem Nokomis urges Hiawatha to: "Slay this merciless magician, save the people from the fever that he breathes across the fen-lands and avenge my father's murder!" The "merciless magician" was the mosquito, of course, and this association with fevers predates Ross by fifty years. Unfortunately, in 1855 mankind had not yet figured out that only the female mosquitoes take a blood meal, so even he had the gender wrong.

Not all poets treat insects with such heavy subject matter. Ogden Nash could usually be counted on to see the lighter side, and he did so with this little rhyme: "Some primal termite knocked on wood; tasted it, and found it good. That is why your Cousin May fell through the parlor floor today." But I especially like what Hans Christian Anderson had to say about insects. "Just living is not enough," said the butterfly. "One must have sunshine, freedom, and a little flower."

It must be nice to be a poet and be able to capture profound thoughts in beautiful language. And when my cold, hard, data doesn't seem to make sense, it would be nice to be able to just claim "poetic license".

JUNE

Week 3

THE WORTH OF AN INSECT

Some people say that this is the "age of man", but that is probably just because men get to say such things. If insects got to write books, they would probably say this is the "age of insects." Estimates from various sources differ, but most experts think the number of insect species is close to a million. However, there are at least twice that many that have not been identified. Some people think there could be as many as thirty million different species. They probably represent about eighty percent of the known animals in the world. At any given time, there are some ten quintillion (10,000,000,000,000,000,000) individual insects alive.

I don't know if mere numbers make insects any more valuable than men, but from a biological perspective their numbers are impressive. They must be doing something right, and they have been doing it for a lot longer than humans. The earliest insects seem to have originated about four hundred million years ago. The earliest human-like critter is estimated to have lived around two hundred thousand years ago. Now I am the first to admit that age doesn't always produce wisdom, just ask my children. However, insects seem to be good at staying alive.

I don't know whether this is the age of man or the age of insects. I am not even totally clear on what age I am. When I recently asked my doctor why I was having trouble with my eyes, I was told that I was old. Then he charged me money for that. So being old, and therefore wise, I still think deciding which animal is more important than another is sort of beyond me. We are all so tangled up together here on this earth. If we want to talk about the value of one animal over another, though, I have a suggestion.

According to the USDA, the value of pollination to

agriculture in the United States was fifteen billion dollars in the year 2000. Then, according to the American Bee Journal, there were approximately two million five hundred thousand beehives in the United States, in 2010. While these data are separated by ten years, they are the best I could find. But that puts the value of a single beehive at close to $600.00 per hive, and that is just counting the hives value to the public in pollination.

According to Bee Culture magazine, the value of honey produced in the United States last year was close to three hundred million dollars. That, divided by the number of hives, adds another $120.00 to the value of the hive for a total of $720.00 per hive. Of course, there is also a market for bees, bees wax, pollen and propolis. Now most beekeepers can only dream of making $700.00 per hive. Like the old saying goes, "There is a lot of money in beekeeping. Most of it is going out."

Then I attended a seminar a couple of weeks ago. The presenter said that Pennsylvania State University had conducted a study and found that one bee hive was worth about $13,000.00 dollars to the public in goods and pollination. (I have to assume their data is better than mine. After all, they are a University.) Of course the study was measuring the entire benefit to the public. Since bees can forage for up to five miles, and routinely forage for three miles, many gardens and commercial famers benefit from pollination they never pay for. That seems to be the big difference between our estimates.

I don't know if that makes bees more important than any other animal. It does sort of indicate that maybe communities and neighborhoods ought to be encouraging beekeeping, not discouraging it. Just for the public good.

JUNE

Week 4

ART AND SCIENCE

Science is concerned with physical things. Science is born of questions such as: How many are there? Why do apples fall down? How does a falling thing fall? What shape is it? How big is it? How much does it weigh? Why does that object act that way? This requires scientists to restrict their attention to a single object or event and study that one object or event carefully.

This study may require physical skills and special techniques. The scientists may even have to invent new methods and perfect new skills to conduct their studies. Often studies are done which simply attempt to establish a pattern or direction. But from this careful, and sometimes lengthy process scientists attempt to distil some kind of general understanding about the objects or events that they have studied.

This general understanding is sometimes called a theory. As it becomes more reliable and useful, it is sometimes called a law. These general ideas can then be used to compare other similar objects, evaluate the theory further, and make predictions about events under certain conditions.

But, overall it appears that scientists begin with some real-world physical object or phenomenon and conclude with a general idea. They turn the world of reality into the world of imagination and thought.

In contrast, art appears to be concerned with ideas. Much of art, including visual art, music, language arts and performance, appears to be born from such matters as: religious concepts, political movements, cultural characteristics, imaginary events, or social ideals. This requires artists to restrict their attention and focus on a specific idea they wish to explore.

To produce pieces of artwork requires artists' physical skills and sometimes special techniques. Artists may have to invent new methods and invent new skills to conduct their studies. Often artists make several models, or attempts, to capture the ideas being contemplated into tangible forms.

In the end, artists create physical objects which represents their views of an idea. The end product of art are functions of the physical world. They may be visual, audible, or palpable; but they are real. These objects can then be used to test the accuracy of the artists, and societies, understanding of the ideas, explore the ramifications of the ideas, explain the ideas more fully to others, or even test the truthfulness of the ideas.

But the overall conclusion is that artists tend to begin with some non-physical ideas and conclude with real objects or physical manifestations that can be detected by the senses. They turn the imaginary world of ideas into reality.

Thus, it seems that both scientists and artists are concerned with understanding our world, arriving at some form of truth and increasing understanding. Both utilize existing knowledge, personal skills and equipment. What appears significantly different is that they initiate their mental journeys from separate starting points.

Because of their opposite trajectories, scientists and artists often see themselves as in "conflict". But understanding the similarities of the two endeavors enriches each field significantly. This can be especially powerful in educational endeavors where numerous studies and pilot projects have shown that using one approach to study the other is especially effective.

For example, having students write about math or science has increased understanding for many students. Writing computer programs that artistically animate scientific phenomenon has proven animation to be an excellent teaching and learning tool. The discipline of assigning artists to explore specific scientific concepts in art classes leads to greater understanding of both art and science.

I think the world appears to need fewer engineers and poets. We need far more people who understand the relationship between ideas and objects. The creation of ideas has an effect on the physical world. The creation of objects has an effect on the creation of ideas.

JUNE

Week 5

(There ought to be five weeks in June, even if there aren't.)

JELL-O

When I was late for dinner as a boy, my Mom put food on my plate and set it on the table until I came home to eat. I didn't mind cold food too much. But the hot food on the plate always melted the Jell-O which would then run all over my plate. I hate cherry Jell-O in peas!

Gelatin is an interesting compound. It is derived from a protein in vertebrate animals, called collagen, which is found in cartilage and tendons. It is used as a thickening agent in foods, pharmaceuticals, and cosmetics. Most of us recognize it as Jell-O. Gelatin thickens as it cools but turns watery at room temperature.

One hundred and fifty years ago, or so, humans discovered bacterial diseases. An early problem with growing bacteria was that they could only be grown in liquids. That meant the bacteria were all mixed together. It was difficult to get pure cultures of just one kind.

Then a German physician named Robert Koch noticed some colored spots growing on the surface of potato leftovers. Refrigeration hadn't been developed yet. He teased a little off with a needle and looked at it under a microscope. It was bacteria, of course, and they all looked alike. It struck him that maybe it was a pure culture and that using a potato surface could help him grow pure cultures of bacteria.

Growing bacteria on a surface kept the organisms in one place. This turned out to be the key to isolating and growing pure cultures of bacteria. It was a major breakthrough in bacteriology. However, there were difficulties. Some bacteria liked potato about as much as I liked cherry-Jell-O-peas. Who knew bacteria were such fussy eaters? So in search of a better surface area, Koch tried different gelling compounds, including gelatin.

Bacteria grew on gelatin surfaces just fine. Only gelatin had to be kept chilly to have a surface, and many bacteria don't grow well when it is cool. You wouldn't think a single celled creature could be so selective in its habits. By contrast, no one pays much attention to what temperature I think my office should be. But when it got warm in the lab, the gelatin would turn to soup. It then ran all over the plate mixing up the bacteria again.

Koch had an assistant in his lab named Walter Hesse. Hesse's wife, Angelina Fannie Hesse, also worked there part-time. Their gelatin kept melting in the summer heat, and they were frustrated. One day, over lunch, Walter asked "Lina" why her jellies and puddings stayed solid even in the summer heat. She told him about a heat resistant gelling agent known as agar that she had learned about as a child growing up in New York. She had learned about it from a Dutch neighbor who had emigrated from Java and such is the nature of science. An Asian women taught a Dutch immigrant about agar. She in turn taught an American girl in New York who married a German, and this German brought it into medical research.

Agar is derived from seaweed and has some unusual properties. Agar is a polysaccharide that doesn't melt until above 100°C, but then, remains a liquid until it cools below 50°C. Once it has gelled, it remains a solid until it is heated above 100°C again. That means it is a solid at room temperature and is perfect for growing bacteria on its surface.

It is found in many foods and consumer products. It keeps stuff from running into the peas, I suppose. You might be interested in checking out the contents of some products in the grocery store for words like agar, carrageenan which is a similar seaweed product, agarose, agarbiose, D-galactose, or 3,6-anhydro-L-galactopyranose. OK, forget that last one. I can't remember it either. But here's a hint. Look in toothpastes, vitamins, ice cream, antacids, jellies, noodles, soups, sauces, and candy bars.

JULY

Themes:
- Freedom
- Fireworks
- Sunburn
- Heat
- Ice Cream

If I lived someplace else, I might have a hard time thinking of what to write about in July. But being an American, the whole month is about freedom, fireworks, sunburns, heat, and ice cream. You might not think that freedom and fireworks are scientific, but you might be surprised. You will probably at least be surprised at what I call scientific.

I hope no one is offended by my take on scientific freedom or weather and sunburns. You'd be amazed how some people don't want anyone to write anything they don't agree with. The worst seem to be "climate change" types. Since I will be talking heat and light, I hope they aren't unhappy. Atheists are almost as bad as the climate change people, but this month should be alright with them. Oh, wait! Freedom has some religious aspects to it too. Oh, boy!

JULY

Week 1

THE SCIENCE OF FREEDOM

Science is the study of the material world. The material world works according to a series of physical laws. These laws seem to have been universal throughout time. The laws also seem to apply everywhere throughout the universe. Where these laws come from, or why they exist, has never been adequately explained by science. In a universe emanating from an infinite explosion, one might expect a world of chaos. Why there are order and physical laws is a mystery to science.

The material world of science raises certain questions for those who study human philosophy. We are made of material, and so our bodies obey material laws. Therefore, some scientists have come to the conclusion that all of our actions stem from the interactions of these materials and physical laws.

These scientists say that human behavior is the result of chemical and physical reactions in our brains. Human behavior becomes nothing more than what we have been "preprogrammed to do" since the beginning of our universe. Some scientists claim that we are simply the results of our DNA and the body's chemical ability to reproduce itself.

The question then becomes one of mans' free will. Many well-known scientists have specifically stated that free will does not exist. Francis Crick says, "It seems free to you, but it's the result of things you are not aware of." E. O. Wilson states, "The hidden preparation of mental activity gives the illusion of free will." Doesn't it seem strange that these scientists, who specialize in the material world, fall back on the description of invisible activities to explain reality? They never let theologians do that.

Crick and Wilson were under no physical law, or threat, compelling them to write those words. I feel no compulsion to write similar words. Do they think those words help them fulfill the destiny of their selfish genes? If the physical laws governing biological systems compel them to propagate, why have both *chosen* to remain childless? Their choices seems contradictory to their claims that there is no free will.

As I sit and write this chapter, I have a sandwich next to me on the desk. I can stop and take a bite, or I can ignore the sandwich and let it grow stale. I could throw it across the room if I wanted to. Once I've tossed the sandwich, physical laws will take over and determine its direction and rate of fall. But the decision to throw it is mine alone. So the action would not be the result of physical laws, but of my own free will.

If free will does not exist, then the entire vocabulary concerning praise and blame, approval and disapproval, admiration and contempt is eradicated. There is no longer any "should" or "ought" in the decisions we make. We could go so far as to say that we have no grounds on which to condemn any crime. However, I bet if I punched Richard Dawkins in the nose, he would say that I shouldn't have done it.

Scientists such as Richard Dawkins, Francis Crick, and E. O Wilson believe that humans are no different from animals and occupy no special place in creation. Yet by their own words and actions, they acknowledge that there is one creature, of all the creatures, that can deny their genetic demands and alter the physical laws governing their brains. And the only creatures capable of these things are humans.

As humans we are, apparently, not totally free. I am not free to alter the course of the sandwich through my digestive system. (I didn't throw it. It was peanut butter and honey for

Pete's sake!) But, on the other hand, I am free to offer charity to a man on the street. The actions of my body are subject to physical laws, but the decision to offer gifts is mine alone.

It is fascinating to me that the entire universe appears to be governed by physical law, but only one creation can make decisions that can alter the course of these laws: mankind. Freedom!

JULY

Week 2

LANGUAGE AND FIREWORKS

One of the cool things about becoming a scientist is that you get to learn all kinds of fancy words that you can use to sound intelligent, whether you are or not. Most of the words mean the same thing as normal words, but sound more cool.

Before we get into that, did you know that if you carefully heat bamboo sticks, they will explode with a loud bang? That is because the gas trapped inside each segment expands until it ruptures the tough fibrous covering. I am told that Chinese children used to do this believing that the loud noise would frighten away monsters. Sort of a poor-man's fireworks, I guess.

The Chinese invented gunpowder about thirteen centuries ago. Interestingly, gunpowder wasn't originally used for guns. It was used for fireworks. And fireworks were used in ancient China, as they still are today, for entertainment, celebration, and to scare away evil spirits. These celebrations weren't for the common folks, though, just for the rich and royal. China is still the leading manufacturer of fireworks in the world.

The way fireworks work is by igniting small pieces of burnable material called stars. Stars are pellets of various metals, salts, or other compounds that have specific colors or effects when they burn. A wet paste is made of the burnable material. Then this paste is turned into small pellets through a variety of simple methods. The pellets are the "stars".

Since many of the compounds are difficult to light, the stars are then coated with a primer material that ignites them more easily. Often black gunpowder is used. The stars are packaged into cardboard or paper tubes and fitted out to either ignite the stars on the ground or be projected into the sky for ignition.

This brings us to the part about big words. Each chemical compound burns with a slightly-different-colored flame. When you see the color of the flame, you can make an observation about the contents. (You might be entirely wrong, but no one will know, unless you're unlucky enough to be close to a science geek of some kind. However, science geeks are pretty rare, so the odds are good you can pull this off!)

So, you don't have to take a course in chemistry to enrich your life with chemical knowledge and sound intelligent. Just follow this basic primer to enrich your conversation and impress your friends this July 4th. First, you can scatter historical comments about China and history throughout the evening, and then use your new-found chemical jargon during the finale.

Instead of the usual "ooohs" and "ahhs", you can learn the following few names, to be used appropriately. If your memory is poor, or time is lacking, you might be able to fill out a little cheat sheet that could be consulted while others gaze at the sky.

For example, if a particular explosion has an intense blue color you could observe, "Wow, those copper halides are brilliant!" Since there is often more than one blue firework throughout the evening, you could vary this with, "Oooh, I think that was a copper chloride."

If a display is more of an indigo hue, you can observe, "I wonder if that was cesium or potassium?" At a shower of gold sparkles you can say, "You realize, of course, that is mostly just iron." Strontium and lithium both burn red, although strontium is a more intense hue, so you could speculate about which you think each burst might contain. Those really bright, white explosions are usually fueled by aluminum, or a mixture of aluminum and magnesium called magnalium. During a lull you can explain that magnesium is too dangerous to use, so they have to form a more stable mixture using one of the other two elements.

Green flames and stars are most likely made using barium. And "bury him" is what many of your friends will be tempted to do to you by the end of the evening.

JULY

Week 3

SUNBURN

There are so many problems in the world today. Let me shed some light on one of them. The problem upon which I wish to shed light is sunburn . . . and why would I want to shed light on sunburn? Thanks for asking!

Sunlight has a wide spectrum of wavelengths. Included are very short wavelengths called ultraviolet, or UV light. UV light is a form of energy that can cause damage to cells. If too much damage is caused, the cell is killed, and we call that a sunburn. When UV light reaches the DNA in skin cells, it forces two sides of the DNA molecule together causing a distortion of the DNA molecule. This creates a break on each side of the molecule.

However, in a cellular version of "If-you-break-it-you-fix-it", many of the breaks are repaired by a mechanism called light repair, or photo-reactivation. While UV light is breaking down the DNA, the visible light portion of the light spectrum causes your cells to produce an enzyme called photolyase. Photolyase repairs DNA by tearing open the misshapen, damaged area of the DNA and reforming it into its original, undamaged shape. (It's all imaginary. So you can imagine little mechanics with wrenches running around in the cell fixing broken things. That makes it easier for me.)

So, even though you break DNA down with sunlight, light also triggers the repair mechanism that fixes the breaks. The health of the cell becomes a race between breaking and repairing. Of course, a break has to occur before a repair can be initiated, so repair is always lagging a little behind. Over short periods of time, the repair mechanism can keep up. But the longer the cells exposure to UV light, the more breaks occur, and the fewer breaks are repaired.

This process mirrors my trucks history. The truck breaks down. I repair it. Something else breaks again, but now I'm broke so the repairs have to wait. While I'm waiting something else breaks again, and again. Pretty soon the trucks mostly broken. In the same way, over a period of years, exposure to sunlight is accumulative. The accumulated damage done by UV light-induced breaks becomes mutant cells that are a form of cancer.

Recent research at Ohio State University by Professors Dongping Zhong and Robert Smith, has shown that photolyase doesn't repair the breaks on both sides at once. It's a two-step process that sends an electron across the DNA molecule in a circular route from one breakup site to the other. This all happens very fast, over extremely short distances, but they were able to measure the time difference in the two-step process. This was done by using a strobe laser that can take measurements in time periods of 90-trillionths of a second. (And you thought those kinds of numbers were only useful in politics.)

Theoretically, the electron could jump straight across the DNA molecule to repair the cell. However, there is another molecule the electron follows that acts like a bridge. So even though it covers a slightly longer distance, the electron travels more quickly. You can compare it to the city center bypass.

So the way you fix UV, light-induced breaks in DNA is by shining even more light on the cell, albeit light waves of the proper wavelength. And the way we measured the two step process is by using a strobe light to capture changes over short periods of time.

"There are so many problems in the world today. Let me shed some light on one of them." Get it? We get the breaks when we shed light on cells. We repair the breaks when we shed light on them. And we measure the repair process by shedding light on it. It's been said that the pun is the lowest form of humor. I suppose that makes this article an even lower form of humor than usual, for puns that is. Does this mean I get a trophy or something?

JULY

Week 4

COOL CAT ON A HOT TIN ROOF

Some of my garden potatoes are "done" already. Yeah, in this heat, all we have to do is pull them out of the ground and add a little butter and salt. There are extra expenses though. My son had to start feeding his chickens crushed ice so they wouldn't lay hard-boiled eggs.

Heat doesn't affect me much. I know others complain of lethargy and mental confusion. Me? I'm just as ambitious and clear headed as I have ever been. My wife says that is not saying much.

I think I stay cool because I have a great circulatory system. Occasionally my cholesterol gets low. But I just get some hard-boiled eggs from my son's chickens, and I'm good to go. (Pay no attention to Dr. Sullivan over there. My circulation is fine.) You probably thought your circulatory system was for circulating blood. Well, sure, but it also circulates oxygen, carbon dioxide, nutrients, hormones, urea, and . . . heat.

We humans are fastidious in our temperature requirements, although not much else. If our body temperature gets much below 96^0 F, the chemical reactions required for maintaining the body don't happen or happen too slowly to help us keep up. Above 102^0 F, the proteins that help regulate our chemical reactions break down. The breakdown of proteins from heat is normally called cooking. In either case, six degrees is a relatively narrow range of temperatures compared to other living things.

High temperatures can result in red faces, swollen hands and feet, and a variety of other problems. The body core is a lot warmer than the extremities. So the circulatory system moves heat around by constricting some of the arteries or veins and relaxing others. The heart rate may also increase to speed the cooling process.

If internal temperatures get too hot, blood vessels leading to the skin and extremities like the ears, nose, fingers and toes will relax, increasing blood flow to these areas. The extremities have larger surface areas and radiate a lot of heat which cools the body.

Like working with the government, there are usually unintended consequences. The more blood being sent to the toes, the less blood there is to send to other organs, like the brain, so other blood vessels must constrict. Less blood to the brain not only keeps you from being a "hot head", but it also reduces the amount of sugar and oxygen available for normal thought.

One of the first symptoms of decreased blood flow to the brain is increased irritability, which can certainly be irritating. Irritability is sometimes followed by loss of concentration, mental deficiency, loss of effectiveness in skilled tasks, depression, confusion, and paranoia. In extreme cases there is obsessive compulsive behavior concerning ice cream, and a gradual descent into madness. These more severe symptoms don't usually occur until it's been hot for at least a week or so though, so you don't need to worry.

Sometimes surface area isn't enough to keep the body cool. Then the body helps out by sweating. The water from sweat glands evaporates and helps cool the skin which is made hot by the excess of hot blood. This is cool as far as it goes. But if it goes too far, it can lead to a loss of fluid from the body. Blood is mostly water. So when fluid is lost, there is even less blood to send around.

Our bodies may decide that using muscle, which creates yet more heat, is not a very good idea. Therefore blood flow to the muscles is reduced. If the blood supply is reduced to the digestive organs, appetite may be suppressed, or indigestion may occur.

    I do wish it would rain though. I don't wish it so much for me as for my six year old grandchild. I mean, I've seen it. Like I said, heat doesn't affect me much though. I'm just as ambitious and clearheaded as ever.

JULY

Week 5
(Only if there is air conditioning)

A QUART OF LIGHT

We don't usually think in terms of a "volume of light". We tend to speak of brightness. But the amount of anything available in a given space is called a volume. "Please pass a quart of light," sounds strange doesn't it?

But think of it this way. If you focus light with a lens on a given point, it will form a cone. Anyone who has ever eaten an ice cream cone knows that a cone has a volume and that one can only increase the volume by increasing the height of the ice cream. The diameter is fixed by the physical diameter of the opening to the cone.

So if you want to know exactly how much an ice cream cone will hold, you can calculate the volume of a cone with the formula: volume = 1/3(Area of Base)(Height), or 1/3 (pi x r^2) (h). The opening to a cone is called its base and is, by definition of a cone, circular. So you have to first calculate the area of the circular base.

The distance around a circle is called its circumference and the distance across the circle through its center is called its diameter. I am sure all of you recall that if we divide the circumference of a circle by its diameter, we get the same number every time, called pi. You don't remember? For Pete's sake, we covered that in the eighth grade! Well, anyway, the number pi is approximately 3.14. (Does it make you feel any better to know that I couldn't recall any of this either and had to look it up on Google? Why remember stuff you can always look up?)

The reason you need to know things like diameter and pi is so you will able to calculate the area of the base of your cone. Then you will be able to select the cone that holds the most ice cream. Knowledge is power! But wait! We aren't done. To calculate the area of the base, you have to multiply pi times the radius. The radius is one half the length of the diameter. The height part is pretty simple though. You just measure that. Then you multiply the area times the height, and then multiply that number times one third. Don't ask me why the one third part, but it works.

Of course, no one seems to make the pointy ice cream cones anymore, and that messes everything up. But, when you focus light on a point it still makes a cone. So you can still calculate the volume of light in your cone. For example, if you had a cone with a diameter of 20 inches, the area of the base would be 314 square inches. If your cone was forty-eight inches high, the volume would be between about 5000 and 6000 cubic inches, depending on how accurately you measure everything. (I know, scientists usually measure in millimeters, but this chapter is for real people.) And if inches are too difficult, twenty inches is about the length of most people's forearm from elbow to the tip of your fingers.

If you have 5000 cubic inches of light for several hours, you can probably cook food with a cone shaped solar cooker. If you make the base wider or the height taller, you will get more light and, hence, more heat. But an increased volume of anything also takes longer to heat up, so it is a tradeoff. The above dimensions work pretty well.

The advantage of using a cone is that, if there is a reflective surface on it like tinfoil or something, the foil tends to focus the light into the tip to give a higher temperature at the focal point for better cooking.

Of course, if you are more interested in ice cream, you probably shouldn't put it into this kind of cone. With ice cream, it is probably easier to change the height of the ice cream than to change the diameter. I'll have three scoops, please. But for my baked potatoes, please pass eighty seven quarts of light.

AUGUST

Themes:
- It's Hot
- It's Humid
- Rock Skipping
- Dog Days
- Sunshine
- It's Normal

Let's head for the mountains and skip rocks on a lake.

August is almost as bad as January as a "do nothing" month. Weeds grow like crazy, so there is no sense in trying to keep up with them. The bees are already starting to think about winter because there is a nectar dearth. It's terribly hot, and humid. Well, it's normal for August.

AUGUST

Week 1

HELL

WARNING: This essay contains a discussion of a common profanity and may not be suitable for all readers.

Last winter, a friend observed that it was "colder than Hell". Recently, I overheard comments about it being "hotter than Hell." I corrected my friends' language on several accounts, but let the more recent comment go. I didn't know the person, and we live in dangerous times. But I pointed out to my friend that Hell is generally assumed to be on fire, not cold at all, and that "colder than hell" didn't make any sense. My physicist friend and I then had a discussion concerning Hell that I thought might be enlightening to the general public.

It's a little hard to get good data on Hell. There is a lot of speculation about who's going there and such, but hard data is lacking. However, one source that should be considered reliable is the Bible. Revelations 21:8 refers to Hell as a lake of fire and brimstone. Brimstone is an old fashioned name for sulfur. If people are to be in a lake of sulfur, then the temperature must be high enough to melt the sulfur, but not so high that the sulfur becomes gaseous. Sulfur changes from solid to liquid at about 115 degree C, and from liquid to gas at about 445 degrees C. So Hell must be a temperature somewhere between these two temperatures.

Anyway, as you can see, being "colder than Hell" as my friend claimed it was, is not really any great accomplishment since 115 degree C is higher than boiling water. However, I suppose he was technically correct since it very definitely was "colder than Hell" at the time of his comment.

In fact, it is almost always "colder than Hell" on earth because earth temperatures do not often exceed 115 degrees C. It isn't even a particularly rare event on earth when "Hell freezes over". I suppose it is frozen over most of the time, except possibly on a few rare days each summer in Arizona, Death Valley, or wherever.

However, when they say it is "hotter than Hell" down in Arizona, I think it is usually just another exaggeration. I doubt it is ever really "hotter than Hell" on earth. I don't know what the record temperature on earth might be, but I doubt it is above 445 degrees C, which is the temperature at which the lake of brimstone would cease being a lake and would become a cloud. I don't know what you would call Hell if it actually got "hotter than Hell" in Hell.

But that brings up a confusing issue. Temperature and pressure are always related. There are a whole bunch of laws about this: like Boyle's Law, Charles's Law, Gay-Lussac's Law, Avogadro's Law, and, well, you get the picture.

But basically these laws all mean that if the volume decreases, or the pressure increases, the temperature increases. So the temperature of hell sort of depends on how many souls are being added and the actual volume of Hell. Since people have apparently been going to Hell since at least the time of Dante, hell must be infinitely big. But in that case, there would be no pressure, so the temperature would be zero degrees Kelvin. Or it must be a set size, but expandable, so as to keep the temperature between the two temperatures of 115 degree and 445 degree C.

If Hell is gaining souls faster than it can expand (which seems likely in an election year), then pressure is increasing, and increased pressure causes increased temperature. If the pressure increased too much, there would be one Hell of an explosion. On the other hand, if there were a world-wide revival and massive repentance during an off-election year, the number of new souls arriving in Hell might decrease, thus causing the pressure and temperature to drop. That is when Hell would freeze over.

AUGUST

Week 2

ROCK SKIPPING

It's hot, and we're heading up to the mountains to skip rocks on a lake. I suppose we could skip rocks down here, but it's cooler up there. I know a lot of people go fishing in the mountains, but skipping rocks is a lot cheaper and more scientific. That's not to suggest that fishermen aren't scientific. I just don't usually trust their data.

The world record for stone skipping is fifty-one skips and it is held by Russell "Rock Bottom" Byars of Pennsylvania. He just did this on July19, 2012, at the junction of the Allegheny River and French Creek about 70 miles north of Pittsburgh. Guinness World Records experts analyzed film of his throws, checking the concentric circles in the water, by each skip. Apparently they didn't trust his data either.

Skipping rocks is the ultimate summer activity although it is a lot easier to do in the winter when the water is frozen. The rocks are harder to come by in winter though. It seems they are either covered with snow or frozen in the ground.

The traditional rock for skipping is one that is relatively flat on one side and mostly round. These are thrown sidearm, parallel with the water, and the rock is kind of rolled off the tip of the index finger, so it is spinning as it flies.

When the rock strikes the water, it displaces the water and causes a concavity, or depression, in the water. The back end of the depression is less dense than the front end so the rock slides down the depression on its flat side. Then the rock glides up the up-side of the depression, because the water is denser there, and exits the water. It's like a skateboarder sliding down a curl and then shooting up the other side.

Did you know you can skip round rocks, not just the typical flat ones? The best rocks for this are very round, or a little oblong. They have to be thrown very hard, overhand at the water, and at about a forty five degree angle. As it rolls off your fingertips, put some back spin on it.

The back spin will create lift, somewhat like a swimmers stroke that is partially used to keep the swimmer on top of the water. A rock thrown this way will disappear under the water and resurface a distance away. Your friends will be amazed.

You might think this chapter is trivial, and you're probably right. However, knowing how solids interact with water has some important practical and scientific considerations. For example, the fields of transportation, energy, and medicine all have a concern with the interaction of solids with liquids.

This subject is so important, in fact, that Jan von Heland of Sweden, invented the "Water Bouncing Ball", or "Waboba". Apparently he grew tired of skipping rocks on ice and decided to invent a better water skipper. His invention is a silicon gel-filled ball that is water proof and floats. This is a great invention because you never lose your best water skipper like you do with rocks.

The secret of the Waboba is that it is flexible, kind of like a hacky sack. Even though it is round, when it strikes the water, it flattens out like a flat stone. Because it flattens every time it strikes the water, it tends to skip better than rigid stones. You can literally play catch with a Waboba by bouncing it off the water.

They really go if you shoot them out of some kind of air canon. There are a lot of directions on the internet for making these, but I am not going to post them in the interest of public safety.

There isn't much of anything better to do, when starting a young person off on a path to science and math, than skipping rocks. Besides, eventually you almost always fall in the water and get wet. Isn't that the point?

AUGUST

Week 3

THE "DOG DAYS" OF SUMMER

It's a dog's life. So why do dogs get blamed for the hottest month of the year? I know, their tongues hang out and they look hot. I do the same thing. Dogs' tongues hang out because they can't sweat. They have very few sweat glands in their skin, and most of those are located in their footpads. Did you know dogs have sweaty feet? You probably already knew that dogs cool themselves by panting. By breathing rapidly, they use the moist surface of their lungs to evaporate moisture and cool themselves.

Water has what is called a high "specific heat". This means that it takes a lot of energy to raise the temperature of water one degree. So when water evaporates, it gives up that same amount of energy and cools the surface. Dogs produce thin, watery saliva that evaporates easily on their tongues, and this also helps cool them.

That is <u>not</u> why August is sometimes called the "dog days" of summer though. The ancient Romans noticed that during the hottest part of the year the brightest star rose in the sky about the same time as the sun. This star, named Sirius, is the prominent star in the constellation *Canis Major* and so the star is also called the Dog Star. (It isn't by chance that an important character in the Harry Potter series is named Sirius and is an *Animagus* as a large black dog.) The Romans thought that it was the combination of the sun and Dog Star that caused the heat, so they called the hottest days of summer the "dog days" of summer. We've been calling it that for at least two thousand years now.

So why don't earths' hottest days occur when our planet is oriented almost directly at the sun instead of two months later? This seems counterintuitive. The sun is farthest from the earth at the summer solstice in June, but it is most directly aligned because of the earths' axis. When the sun is most directly overhead, earth receives the most radiation from the sun because the atmosphere is thinnest relative to the angle of the sun.

This increased solar radiation warms the earth. As the temperature of the earth rises, it radiates energy back into space. However, the molecules of earths' atmosphere interfere with the transfer of energy both in and out. So the energy being lost is partly trapped by the atmosphere. The earth loses heat more slowly than it gains it.

That means the earth is constantly receiving and losing radiation. As the earth warms in June, it doesn't lose all of the heat it gains, but continues to gain heat from the sun. When incoming and outgoing radiation reaches equilibrium, maximum temperatures occur, a couple of months later than the summer solstice and in August.

You can test the heat loss of the earth even in summer. Set a bottle of water out on a clear night, on a cone shaped reflecting surface. I use a cardboard cone covered with tinfoil. Point the open cone towards the empty night sky. In the morning the water will be much colder than the ambient temperature. This is sort of a poor man's refrigerator. The warmth of the water is radiated out into deep space, at the same time cooling the water.

In winter, we receive the least radiation from the sun when we are at the greatest angle from the sun in December. But January and February are the coldest months. The earth loses heat faster than it gains it during the months following the winter solstice.

I am told the North American Indians called the January moon the "Wolf Moon" because that was when the wolves were hungry, cold and howled at night. Did the cold make them howl, or did the howling make it cold? Looks to me like dogs get blamed at both ends of the calendar. It's a dog's life.

AUGUST

Week 4

THAT LUCKY OLD SUN

It's sure hot! I haven't seen any evidence that the sun is cooling yet. However, I am told that it only a matter of time before it burns out. How long that will take depends on how fast it burns. How long it burns depends on what it burns, and how it burns.

So, I got to wondering what the sun burns, how much of it there is, and how the sun burns it. In spite of the fact that all plants depend on the sun, and all animals depend on plants, this hadn't really been a part of my biological education. But I have been asking around to some friends. I couldn't find any friends though so I checked Wikipedia.

The sun burns Hydrogen, a gas, at its core. Burning turns it into Helium, another gas. This process is called nuclear fusion, and it produces heat and light. Someone has calculated that the sun burns about 700 billion tons of Hydrogen a second. Even though that number is less than our national debt, I still found it pretty alarming until I wrote out the number for the mass of the sun, which is 1989 with thirty zeros behind it. That is one million nine thousand, eight hundred and nine billion, billion, billion Kg. Heck, if that were our GNP, we could borrow a whole lot more. Several hundred billion tons of Hydrogen just isn't all that much.

So while it's nice to know that our sun won't be going out soon, it also means that things probably aren't going to cool off a lot in the near future. In fact, some scientists estimate that we only have about a billion years before the sun gets so hot it burns off all the water on earth, putting an end to Lake Powell and the back yard pool. It isn't the heat as much as the humidity.

What I do find kind of alarming is that when the hydrogen is finally all burned up, the sun will begin to burn the Helium. At this point it is supposed to get even hotter until expands in size to encompass the earth, and maybe Mars. So the Sun, presently called a Main Sequence Star, will boil off our water in just a billion years, will last for another five billion years as it is, then turn into a Red Giant for a few billion years, and finally will become a White Dwarf for about ten billion years.

Apparently the sun has already been burning for quite a while so at its present rate of burning, should last another five billion years. I guess it's time to break down and buy the air conditioner my wife has been wanting.

AUGUST

Week 5
(Has school started yet? Please!)

NORMAL

The question has come up as to whether or not I am normal. I immediately became defensive when confronted with the question. Of course, I'm normal! However, I almost always resent being told that I am just like someone else too. It seems I want to be unique, but not too unique, I guess. Does anyone else feel this way?

What is normal anyway? Is normal conforming to social norms of behavior? Is it good to conform to social expectations? Does it make me happier and more successful? Or is conformity a form of tyranny? Maybe I will be happier and more successful if I refuse to conform. If I don't conform to social expectations, how far can I un-conform before society conforms me? It's a fine line.

Or is normal a measure of some average based on a group? It is normal for a healthy human to be able to run long distances, like even ten miles. There is nothing in the anatomy or physiology of most people to prohibit that. Yet the fact that I probably can't run two blocks makes me about average; and average, in this case, isn't normal. Now, why do we call the opposite of sanity, insanity? Shouldn't we be in-sanity when we are sane?

When we measure children by their performances on standardized tests, we compare them to what is average, not what is normal. It is normal for children to learn, and to learn very complicated things, early in life. Consider the great intellectual strides made by children in the first five years of life. Adults seem to be satisfied, once they reach school age, if they have learned enough to be above average. Actually, by the very definition of average, half of all people are above average. The frightening thing is that half of everyone we come in contact with is below average. Do you suppose that applies to elected representatives as well?

This concept is actually more important than it sounds. One cannot define abnormal until there is a clear understanding of what normal is or what normal looks like. Understanding normal is perhaps the most disregarded bit of common sense in existence. Abnormal means "not normal". But until we clearly understand normal, we cannot know what the significance of various deviations from the normal might be.

Knowing what normal looks like is crucial in science. We call the "normal", the control. Then we measure some item or event as it exists in relation to the control. By changing one parameter in an experiment, we can see if the results change. We compare the change to what we presume is normal or the control. Not surprisingly, sometimes even scientists fall into the trap of not defining normal before they begin their experiments.

OK, are we running out of water? Is too much carbon dioxide being released into the atmosphere? Is the earth warming? Are there fewer amphibians than there used to be? Are there more natural disasters than in the past? Is the number of species of insects declining? Are new species being created? What can a child learn by the age of four? How far back do we have to look for normal data?

All of these questions have something in common. In order to answer them, we need to know two things: what normal is and what has changed. In many cases, we don't really know the answer to either question. For example, what is the normal temperature of a planet that seems to have been changing and vacillating in temperature since its creation? What is a natural disaster, and how many have occurred in each century? Is there really less water on earth,

or is the water just dirtier and more expensive to clean up?
Is it better to be crazy and know it, or to be sane but doubt it?

So if I am above average and normal, does that make everyone who's normal about average?

SEPTEMBER

Themes:
- School
- Learning
- Insects
- Honey Harvest
- Confusion

For a long time, September has marked the start of school each year. It seems that is changing, and school is becoming more of a year around thing. I don't think that is such a good idea, but no one asked me. Humans now think the brain teaches the body, so they spend all their time teaching the brain. They have forgotten that the body has to first teach the brain.

The bitter-sweet part of September is that it is the beginning of autumn. The fall colors and dying gardens are beautiful and peaceful in a sad sort of way. Even the bees know it's time to quiet down. The sweet part of the bitter-sweet is that September is often when beekeepers harvest their honey.

Another bright spot in the month is that it is a time for beginning to transition into indoor activities like making music. The Grammys always happen in the fall.

SEPTEMBER

Week 1

THE SCIENCE OF THE GRAMMYS

I love it when music, science and popular culture all come together.

The story of how radio was discovered is complicated and involves many players. But in essence, it was discovered in the late 1890's and early 1900's. When the vacuum tube was born in 1906, it set the stage for major advancements. Radio was on its way.

In 1908 Alvin McBurney was born in Oakland, California, but his family later moved to Cleveland, Ohio. He showed early signs of mechanical aptitude and built his first radio when he was just eight years old. He went on to become one of the youngest people to ever receive a ham radio license, W6UK. Alvin was very interested in electronics and had early plans to become an electrical engineer.

On his tenth birthday, he received a banjo and learned to play. Later he branched out to the guitar as well. The early 1900's was an age of bands and orchestras, but the banjo and guitar often lacked the volume to be heard. So, at the age of 16, McBurney used his electronic ability to build an amplifier for his banjo which he later adapted to his guitar as well. Being young however, he did not apply for a patent until years later, so he is not generally recognized as the inventor of the electric guitar.

In 1927 Alvin landed a job playing banjo with a Cleveland band. Shortly thereafter he got a job playing amplified guitar in an orchestra in New York. Over the next several years, he moved to California, played in several bands, and studied guitar performance under numerous tutors. In 1929 there was a great craze for Latin music, so Alvin McBurney changed his name to Alvino Rey. Alvin went on to become one of the highest-paid sidemen in the field of music of the 1930' and 40's.

Besides inventing the amplified guitar, he developed the steel pedal guitar and developed the prototype pickup for the first Gibson electric guitar, the Gibson ES-150. Just before the war, he also began to use a special mike developed for military pilots. It was called a carbon throat mike and he used it to modulate his guitar sounds. His wife stood behind a curtain and sang along with the guitar lines. This was probably the first talk-box experiment and was dubbed the "singing guitar". Then, during the war Rey went into the Navy and worked on developing radar systems.

Rey met Luise, his wife, when they were both performing in the same band. Luise sang with her sisters, Maxine and Alyce, and they were known as the King Sisters. The King Sisters and Alvino Rey were a group for many years and had numerous hit records. They had a television show for five years in the fifties on ABC. They also had two children: Robert "Robbie" Karleton Rey born in 1946, and Liza Luise Rey, born in 1947.

Liza Rey later married a man named Butler and they eventually moved to Canada. Their two sons, the grandsons of Alvino Rey and Luise King, named Win and William Butler, are members of the Montreal based Arcade Fire which just won a Grammy for Album of the Year in 2011.

I have never watched the Grammies. Of course I've never watched the Oscars either, or the Super Bowl. I don't really watch a lot of television. But I read an article the Monday after the Grammys announcing that some group called the Arcade Fire won best album of the year. The name sounded familiar, but I couldn't quite place where I had heard of them before. Then I remembered that it was in the

November issue of QST, the amateur radio magazine, in an article about the influence of Alvino Rey on modern music.

And that summarizes the science, the music, and the culture part of this article. I love it when music, science and popular culture all come together.

SEPTEMEBER

Week 2

BRAIN LOBES

Humans are all confused about thinking. We think "thinking" is something we do when we play chess or solve math problems. Yet computers do both of these activities extremely well. Well, better than me anyway. Yet a computer cannot do many things we do. Computers cannot hit the snooze button, comb hair, fix breakfast, drive to the office, recognize someone we know in another moving car and honk and wave, find a parking place, and then walk to the office.

Let's take a look at the human thinking machine. The human brain appears to consist of five unequal lobes. When we open up the cranium, these lobes are physically identifiable. Using imaging equipment, we can easily see that each lobe has a unified function as well.

The most posterior lobe, in the back of the cranium, is called the occipital lobe. This is a lucky coincidence because it lies under the occipital bone. The neurons in this area are connected to the eyes, and this is where we experience vision.

On the sides of the brain, directly under the temporal bone and basically behind the ear, is a lobe that receives information from neurons in the ear. This is where hearing is experienced. It is another stroke of good luck that it is called the temporal lobe.

On top of the brain is the parietal lobe. It lies under the parietal bone. (You may be getting suspicious that these names are not coincidental at all.) The parietal lobe receives and sends information to and from the skin and body, informing you where your feet are when you can't see them and what to do with each finger on different hands when you play the guitar.

There are also two tiny lobes under the front part of the brain. These are about the size of your little fingers. They are called the olfactory lobes, and they interpret and respond to smell and taste. Human olfactory lobes are kind of puny when compared to the rest of the brain, and the olfactory lobes of many other mammals. The olfactory lobe of a dog fish shark, for example, is three times larger than the rest of its brain altogether. Ours probably doesn't account for ten percent of our brain. We may smell good, but we certainly don't smell good.

There is one other lobe that is located just behind the frontal bone. It is called, all together now, the frontal lobe. The frontal lobe is unique in that it is not directly connected to any of the body senses but is connected to the other lobes of the brain. We describe its function as being abstract thought. An abstraction is anything you cannot hold in your hand, such as love, freedom, democracy, algebraic X, or the color blue. The frontal lobe is quite large in humans as compared to other animals.

Thus we see that the brain is composed of five lobes, of which four are directly required for sensing and responding to the physical world. Four fifths of the brain isn't for thinking at all, but for responding to the continuous flow of data coming in from our surroundings. "Thinking", as we normally think of it, takes place as only a part of the function of a single lobe and is ultimately dependent upon the rest of the brain for its input.

I think this suggests that "thinking" consists of a lot more than most of us think. More importantly, it suggests that clear thinking requires a rich experience with the physical world. Our ability to form abstract thoughts appears to be dependent on information from the other lobes of the brain concerning non-abstract events and objects.

    As proof of this connection between our physical mind and abstract thought, try to express yourself about an abstract idea, without using physical world terminology such as: big ideas, small minds, loud colors, bright ideas, weighty opinions , heavy news, under handed, looking forward, backward person, tall order, and even the political right and left. Bet you can't say much without these concepts learned through our senses, concerning the physical world. Think about it.

SEPTEMBER

Week 3

INSECTS AND FLOWERS

The insects are dying. Its fall, and it's turning cold. The Honey Bees in my hive are balled up at night now, and there isn't much foraging, even on the sunny days. I see spiders laying their eggs on the side of the house. A Daddy Long Legs, technically not a true spider, hangs lethargically by the front door. The Praying Mantises are big, and fat, and slow. The word Mantis means prophet, and so I assume their size foretells the coming of winter.

The garden is dead. Only wilted and discolored flowers remain in most places. Fruit has been set, seeds have been shed, and nuts are in the shell. I still need to clean out the old growth in the garden, but I don't feel any hurry. Fall is for slowing a little, taking one's time, and feeling a little glad that the work is over. But it's also for feeling a little sad that the growing is over too.

Sometimes the most important truths can be hidden in plain sight. There are over 250,000 flowering plants that have been described. That is probably a modest estimate, but I am not a Botanist and don't want to over-sell. There are over 750,000 insects described. That number is actually much bigger and is expected to go over a million.

Taken together this means that two thirds of all life forms are monopolized by these two groups. This is not an accident. These two groups of living things live together in an intimate way. Flowering plants could not exist without the service of insects to aid them in sexual reproduction, which we call pollination. And most insects could not exist without the shelter, surface, and food (nectar, pollen, leaves, stems, and roots) provided by the plants. These two groups are completely symbiotic: they are dependent on living together.

The concept of living together is a delicate and changing arrangement. There are flowers like *Passiflora incarnata*, the Maypop, common in the southern United States in areas like Tennessee, that are only pollinated by *Xylocopa virginica*, a carpenter bee. If the bee is lost, the flower will also become extinct. Or there is the "bear claw poppy", *Arctomecon humilis*, which is only pollinated by a solitary bee, named *Perdita meconis*, unknown until just a few years ago. If the flower is lost, the bee will go extinct. These last two life forms live near the Virgin River in Southwest Utah, or Northwest Arizona, depending on which direction you're facing.

Sometimes this balance between organisms is upset and we call the result predation, parasitism, disease, extinction, pollution, or some other term. The problem is that it is very difficult to know what will upset the balance between any two or three organisms. How do we know what to avoid, or how to avoid it? The balance is akin to a complex structure built out of toothpicks. It is hard to predict which toothpick can be removed, and which cannot, without causing the collapse of the whole system. Generally, humans don't have a clue as to what we are doing in this regard.

Mankind has put a lot of energy into killing insects. Many insects compete with us for our food. Some insects transmit diseases. But ironically, mankind relies heavily on flowering plants for food and fiber. High mountain peaches, cherries, apples, pears, and apricots are just a few of the hundreds of plants we find desirable, and they rely on insects. So if plants need insects, and insects need plants, and man needs plants, then doesn't man need insects?

SEPTEMBER

Week 4

SWEET MEDICINE

In 1976 my wife and I were involved in a head-on collision that left me semi-scalped from my eyebrows to about half-way back on my head. I looked pretty awful, but it was actually not a serious injury. The doctors sent me home with instructions to apply hydrogen peroxide to the wound several times each day to avoid infection.

Hydrogen peroxide is a compound that has two molecules of hydrogen and two molecules of oxygen. That makes it the same as water, except for its one extra oxygen. When hydrogen peroxide is applied to injured human tissue, the compound is exposed to an enzyme that liberates the extra oxygen, leaving water and a single oxygen floating around. That single oxygen is highly reactive. It attaches itself to any bacteria in the wound, damaging the bacteria's cell membrane, killing the bacteria. The oxygen that is released, however, causes the tissue to foam in a dramatic way.

On the first morning after our accident, my young children were talking to me about what had happened as I applied the hydrogen peroxide to my forehead. They watched in horror as my entire forehead foamed up, a little like a root beer float. They were both fascinated and appalled. Their fascination proved to be the bigger factor as they insisted on being present for every subsequent application. I became the only Dad they had ever heard of with a foaming head. In fact, they asked if they could bring their neighborhood friends over to watch. Sensibly, I did not allow this.

"But what does hydrogen peroxide have to do with sweetness?" you ask. It turns out that honey has the necessary components to produce miniscule amounts of hydrogen peroxide over an extended period of time. Honey is about 30% glucose. But it also contains glucose oxidase, an enzyme from the stomach of bees that is secreted into the honey by the bee. This enzyme, in the presence of oxygen and water, can break glucose down into gluconic acid and hydrogen peroxide.

However, this enzyme does not function in honey because the pH of honey is too low. Honey generally has a pH reading somewhere between 3 and 4.5, and glucose oxidase requires a pH of about 6. Also, for glucose oxidase to function, it requires at least 2300 parts per million (ppm) of sodium to be present. Honey usually has only about 30 ppm. So good clean honey, stored in a proper container, is stable with no reaction occurring.

Human tissues contain an abundance of sodium, and the pH is generally slightly more than 7. If honey is applied to injured human tissue, the pH is slowly raised at the point where the honey comes in contact with the injured skin. The abundance of salt in the body combines to activate the glucose oxidase. This causes the honey to produce minute doses of hydrogen peroxide, over an extended period of time, directly to the place where it may be needed to combat possible infection. However, the honey isn't as fun to watch as the hydrogen peroxide because you miss the foaming part.

Honey is also a supersaturated sugar solution. As such it will not support the growth of bacteria because it pulls the water out of any bacteria present. Honey's low pH also creates an environment that inhibits most bacteria growth. Finally, some honey has been shown to contain anti-bacterial compounds isolated from the floral nectars. In all, honey can be used as a home remedy for dressing wounds.

As you might guess honey varies in its medicinal effectiveness, depending on the floral source of the honey. Other factors such as water content, glucose content, and glucose oxidase content, all play a part. Some honeys, such as Manuka Honey from New Zealand, have greater medicinal properties than others. This lack of uniformity is one reason why honey isn't used more aggressively in regular medical treatment.

Well, that, plus the fact that the honey is far less exciting to watch work than plain hydrogen peroxide!

SEPTEMEBER

Week 5

WHICH WAY IS UP?

I sometimes wonder about how thoughts are originated. Would it have made any difference if we'd thought otherwise?

For example, why do we think that light must come from a source? We all believe that the sun emits something called light. So why didn't we think of the sun as a big dark sucker that sucked up all the dark? Wouldn't that have made just as much sense? There could even be many dark suckers out there, each with varying abilities to suck up the dark. If the world were filled with dark, it would take a really powerful sucker like the sun to suck all the light out. The moon and stars would be less-powerful suckers that couldn't quite get it all.

Perhaps light bulbs also have various abilities at sucking up dark. Within their immediate vicinity, they can pretty much get it all. But a few feet away, they just don't possess enough suction for that dark too. Then when we turn off the sucker, the dark all fills up the space again.

It could work something like this. Maybe there are dark particles called darkons that have something like an electrical charge and are attracted to the opposite charge. What we call light sources would actually be oppositely charged "lightons", and they would attract the darkons. Wouldn't everything still work about the same way if that were thought to be the case?

Here is another one. Why did Newton think there must be an attraction between two bodies that are roughly related to their size? He based his thinking on the principle of acceleration. Since velocity of a falling apple changes from zero as it is hanging on the tree, to something else as it falls, it must have a force accelerating the fall. He imagined that if the apple tree were twice as high, he could expect the apple to be accelerated even more by this force. He then began to realize that this force extended far beyond the apple tree.

So he came up with the law that says, "Every object in the Universe attracts every other object with a force directed along the line of centers of the two objects that is proportional to the product of their masses and inversely proportional to the square of the separation between the two objects." (Its sentences like this that give science a bad name.)

But the question I have is, "Why did he think it was an attractive force?" Why didn't he think, "Every object in the Universe is repulsed by every other object . . . .?" Couldn't the apple have been pushed down from above by the repulsion of the tree mass, or even by the repulsion of heavenly bodies? He could have proposed invisible particles called repulsons, instead of an invisible force called gravity.

To take that thought further, perhaps I actually repulse the earth. (Hmm, that seems to be an unfortunate turn of phrase.) But that would mean the earth also repulses me. Then I'd be being held in place by all the repulsion of the other bodies in space which collectively overwhelm the earth's repulsion for me. Wouldn't everything basically remain the same as far as our experience goes?

We hear a lot about how DNA makes proteins in the living cell. These proteins are essential to the cell because they allow other chemical reactions to occur. Some of the chemical reactions that proteins cause are those that make DNA. We have come to think that life is DNA based. Why do we think DNA makes proteins instead of thinking that proteins make DNA? Further, how did one molecule come into existence without the other molecule already being there, which couldn't have been there because the first wasn't there?

    Don't misunderstand. I don't disagree with the findings of science. I just wonder how we started thinking in one direction instead of the other. I find thinking about thinking kind of weird.

OCTOBER

Themes:
    Daylight Savings Time
    Halloween
    Gore
    Body Parts
    Suffering
    Magic Potions

    I kind of feel bad for October. I mean, almost everything associated with the month is blood, gore, goblins, death, magic potions, and daylight savings. Then it turns out that the saved daylight can't be used later after all. Life used to be so simple, but now there are yard decorations that are almost mandatory, and still the electric bills skyrocket. Read on . . . .

OCTOBER

Week 1

DAYLIGHT SAVINGS TIME

As the shadows lengthen and the sun sinks into the west, darkness gently falls. Quiet settles over the land, interrupted only by the theme song of the O'Reilly Factor. Walking through the streets, you can see each home alight with the loving, flickering, blue-light of television and the occasional steady glow of the computer terminal.

In the inner recesses of each human brain, accompanying the deepening shadows, will be the slow but sure release of N-acetyl-5-methoxytryptamine. This poetic sounding compound, also known as melatonin, is secreted in response to darkness by a tiny, pea-sized organ in the brain called the pineal gland. In humans, the presence of melatonin in the blood causes the lowering of body temperature and the feelings of drowsiness. Ah, sweet peace.

The production of melatonin is inhibited by light. As the sun comes up, or the lights go on, blood melatonin levels decline, temperature rises, and feelings of wakefulness increase. Invigorated and refreshed, we awaken. But basically it's the melatonin levels in the blood stream that help set our circadian cycles of rest and activity.

Melatonin is found in all animals, plants, and microbes. It seems to play several roles, but it's always involved in the timing of periodic events such as sleeping in humans, reproduction in animals that have specific breeding periods, and regulating plant response to day length. All three have something to do with light levels.

However, day length just isn't what it used to be. It used to be that the sun went down, and it got dark. The pineal had a simple job, and it performed like clockwork. But now the pineal has to decide if it's the sun you are staring at or a television screen. Of course one is enlightening, and the other isn't, but it's hard for a gland to know.

Then there's the alarm clock that casts an eerie, red glow all through the night. The speaker on the computer has a brilliantly blue LED, the microwave is yellow and the entertainment center is a sickly golden hue. The street light outside and the night light in the children's room casts shadows down the hall. One hardly notices when the house lights go off for all of these. Since melatonin levels start dropping even in the dim light of sunrise, one wonders if melatonin ever turns completely on in our modern world of artificial lighting.

People who are lacking in this hormone often have difficulty sleeping. Decreased melatonin has even been implicated in premature births, increased number and severity of accidents, and the increased occurrences of some cancers.

But too much melatonin is also not a good thing. In northern climates where the days are short and winter sun indirect, it may never get bright enough to completely turn off the melatonin. Some people suffer from winter tiredness and depression called seasonal effective disorder, SAD, because the melatonin levels stay high enough that they never completely wake up. (This may help explain Alaskan politics.)

But tomorrow, Sunday, November 7, 2010, the gentle light of sundown, the peaceful gathering of the gloaming, and the flickering enlightenment of the daily news shows will occur an hour earlier than it did today as far as your pineal gland is concerned. The gland will be completely caught off guard. Whipsawed by the artificial and unnatural manipulations of ignorant and mortal men, the pineal will vacillate wildly. The majority of Americans will end up having forty minutes less REM sleep on Sunday night, almost completely obliterating the gain of one hour the night before.

On Monday we will awaken to the carnage of an increased number and severity of automobile and industrial accidents, grumpy and lethargic coworkers, and confused pineal glands across the land. It will take a week for all the pineal glands to accommodate to the change that is daylight savings. Each spring it happens again. Leave the blasted clocks alone!

OCTOBER

Week 2

AN EYE FOR AN EYE

Some of the stuff I find scientifically interesting just isn't very practical. How do cats lap up milk with their tongues? Have you ever tried to do that? Why are bees' eyes hairy? The fact that the Universe is expanding surely affects us somehow. It's scientific, but it just doesn't seem very pressing.

Sometimes science is very practical and delivers new, important products like iPhones and Wii consoles. Science has been used to create chickens that are ninety percent chicken breasts. Science has created watermelons the size of grapefruit, but the smaller versions cost the same as real watermelons. See, very practical.

At the same time, there are a lot of practical things that aren't very scientific. Like life and disability insurance. Did you know that if I die, my wife will get ten thousand dollars? That's practical? I'd call it incentive! The fact is I am only worth $10,000.00 because I am so old. If I was younger than twenty five, I would be worth $120,000.00.

Now that kind of reasoning is patently unscientific. Show me a twenty-five-year-old who is worth that kind of money. I was that age once, I know. However, now that I might actually be worth something, the insurance company says I am worth less. Does that sound scientific to you?

In the disability part of the policy, insurance companies put different values on body parts. This practice certainly isn't biblical. "The eye cannot say to the hand, I have no need of thee." If I lose both hands, both feet, both eyes, or one hand and one eye - but I am still alive - I will still get the full ten thousand at dismemberment. Boy, in today's economy that is really tempting. It isn't clear if I have to lose all of those things, or just one of the pairs to collect. Do I get the money if I lose both hands, or do I also have to lose both feet and eyes? That wouldn't be practical. I mean, if it's just one hand and one eye, then it might be worth it.

You would have to be pretty determined to cash in though. If you quit after the first appendage and only took off one hand, one foot, or one eye, you would only get half of the settlement. That makes the other body part equal in value to a whole body. That doesn't seem scientific. Another thing, is half my life over if I lose a leg? This assumes that hands, feet and eyes are all of equal worth and that without all of them you aren't worth anything at all. They may have a point there.

I wonder what the insurance company would do if you found the part you'd lost. I think they should have to pay even if the doctors reattach the appendage. At the least that would help defray the rising costs of medical insurance.

A hand is worth more than a bird in a bush, and it is obviously worth more than a foot. I can't play guitar with my feet. Is the loss of one eye exactly half the loss of losing two eyes? I don't think so! If you lose two eyes you can't see at all. Insurance companies assume that losing two eyes is the same as being dead. I bet blind people don't think so.

How did insurance people arrive at these values? Did they do a scientific study to determine the amount of pain involved? On who? Is that legal? Was the pain suffered in losing a body part exactly half the pain of dying? How do they know? Putting value on body parts doesn't seem either practical or scientific.

I think these are imminently practical questions which science should address. But science costs money and, unfortunately, the folks with the most to gain from the answer to these questions, are the least valuable people.

OCTOBER

Week 3

MOLES

This time of year people have trouble with moles. We are several weeks into the school year, and chemistry students everywhere are concerned about moles. However, I don't think they are interested in saving the moles. I think they would like moles to become extinct.

They were taught in earlier grades that moles were cute, little animals that lived under the ground and that they sometimes caused bumps in the lawn. Technically, moles are neutral, being neither good nor bad. But bumps in one's lawn are as socially unacceptable as water spots on drinking glasses.

This is how we do it in education you know. We don't exactly tell the truth. Since the world can seem complicated, and, for the most part adults are pretty poor tellers, we gloss over parts of the world with lies. In the second grade we correct some of the lies. This usually results in more lies.

Somewhere along the way, we hope we get all the lies corrected and leave students with the truth. I can tell you from experience, though, that the point of truth in this process is longer than ten years of higher education. I just do not intend to pursue it any further because my wife said I can't.

We put off difficult concepts until children are older. For example, we usually let them think of moles as little animals until they start asking grandpa why he has that funny, brown bump on his face. Exactly when do we introduce the concept of a mole as a pigmented, skin blemish? I think that is best left to the parent's judgment. The public education system should not be involved in this at all.

The truth is: a mole is the amount of a substance that contains as many particles as there are atoms in 12 grams of pure carbon-12. This corresponds to 6.02×10^{23} elementary particles of that substance. What?

Society has decided that some time in high school is the appropriate time to spring this radical information on unsuspecting children. I think that what they meant to say is that one mole is the number that indicates the mass (in grams) equal to the atomic weight of the molecule. For example, water has an atomic mass of 18; therefore one mole of water weighs 18 grams, and contains 6.02×10^{23} molecules.

Lorenzo Romano Amedeo Carlo Avogadro was an Italian born in 1776. I know that date from somewhere. He was first a lawyer and then became a chemist. That right there probably explains a lot of the difficulties people have with moles.

Anyway, Avogadro hypothesized that the number of atoms in an amount of material, equal to the atomic weight of that material expressed in grams, would be 6.02×10^{23} for all atoms.

No, this is not to be confused with the calorie count for Avocado Dip though they are similar in that both are very large numbers. An Avogadro's number of "Avocados" would cover the entire earth to the depth of several hundred feet. I wonder how deep it would be if the Avocados were made into dip?

Another way of expressing this law is to say that equal volumes of ideal gasses, at the same temperature and pressure, contain the same number of particles. Of course, problems arise in defining the words "equal", "ideal", "same", "temperature", "pressure" and "is". When these things are considered, Avogadro's law isn't true at all. The law is just another approximation, which is a fancy word for lie. As soon as you think you understand this lie, I will tell you the truth.

At this point, I hope you are a little less confused about moles. To put it simply, they are either vertebrates, skin blemishes, or a number. These are all really pretty easy to tell apart. But if you are still confused, have hope. Oct. 23, from 6:02 am until 6:02 pm, is International Mole Day sponsored by the National Mole Day Foundation. Save the moles!

OCTOBER

Week 4

CLASS II MEDICAL DEVICES

Maggots, leeches, scorpions, toads, newts, bats, lizards, goats, and spider venom: "by the pricking of my thumb something wicked this way comes." Of course the "wicked something" in the quote was a human; Macbeth from Shakespeare's tragedy of the same name.

From my study of animals, I have concluded that the most frightening are humans. All the other kinds are just part of nature, and they often provide us with surprising benefits. Humans, on the other hand, do some really despicable things.

Most people would say that maggots, for example, are disgusting although not necessarily wicked. Blow-flies are a group of large flies that are brightly colored in metallic blues, greens, and coppers and are actually very pretty when examined closely. Blowflies tend to lay their eggs in dead animal tissue. The immature, the Anti-defamation League suggests the change in nomenclature, consume the flesh and, thereby, aid in the decomposition of dead animals.

It is hard to imagine what the world might be like without the help of blow-fly maggots, er, immature. (We'd be knee deep in rotting carcasses I presume. Blow-flies can even be handy in determining how long a living animal has been dead. The flies are cold-blooded animals and develop at a given rate at known temperatures. They also pass through a set sequence of recognizable stages. Therefore, their level of maturity in a corpse can indicate how long the body has been dead. People who can assess this situation are known as Forensic Entomologists, at least in polite company.

Oh, but there is one tiny problem. Some of the flies lay their eggs on healthy, living tissue. The immature literally eat holes in the skin and consume living flesh. This condition can be extremely costly in livestock as it sickens animals and ruins hides. This condition is called myiasis and can also affect humans. By the way, it is unseemly for people who have watched "Nightmare on Elm Street" or "Night of the Living Dead" to get all fastidious on me here for describing the natural life cycle of a blow fly.

Maggots, er, immature of the common, Green Bottle Fly (*Phaenicia sericata*) eat only dead tissue, and I don't know why they are called a Green Bottle Flies. Is it a green fly, or was it found in a green bottle? Maybe it's a green fly that was found in a bottle. This is a serious advantage in some cases. No, I don't mean the fact that it is found in bottles, because it isn't. The advantage is it eats only dead tissue.

It was noted centuries ago, that some maggot-infested wounds (Oh forget it!) of warriors on the battle field healed more quickly and cleaner than other wounds. In the American Civil War Dr. John Zacharias used maggots to remove dead tissue in gangrene cases, as he said, "with eminent satisfaction." Whose satisfaction, I wonder? That's just gross!

The practice of maggot debridement has suffered over the years for aesthetic reasons, but recently it has been resurrected. Perhaps revived would be a better term. Anyway, it seems that with the development of antibiotic-resistant bacteria, wound debridement by surgical maggots has again shown some advantages. Not only does maggot debridement yield a cleaner wound, but there also appear to be other positive factors associated with maggot-infested wounds. Maggots may have growth-stimulating effects on patient tissue as well as antibacterial effects.

Dr. Ronald Sherman, University of California at Irvine, yet another human, has received FDA approval to produce and market surgical maggots. Yes, he grows tissue-infesting maggots in a lab to sell to other humans, called doctors, to purposely embed in wounded patients! The lab-produced, disinfected maggots are called "non-exempt, Class II medical devices with special controls." Immature? Class II medical device? A maggot by any other name is still a maggot. This is what happens when political correctness runs amok.

    I'm telling you, humans can be positively terrifying. "By the pricking of my thumb, something wicked this way comes."

OCTOBER

Week 5
(Halloween candy hangover...)

SUFFERING

I have been thinking about science and suffering.

I became a scientist, in part, because I wanted to help people. The other part was a useless attempt to avoid the draft. Instead, I now find myself administering exams and writing articles, both of which may cause excruciating suffering for some people.

Many scientists seem to worry about suffering. However, the scientists who worry the most are people who have never seemed to suffer very much. Richard Dawkins, Victor Stenger, and other atheist scientists are greatly exercised about suffering although, they, themselves, have university educations and live rather extravagant and indulgent life styles.

I'm not entirely sure what suffering is. Presumably death qualifies, although I really don't know, not having died yet. I suppose pain causes suffering, but pain levels exist on some kind of a continuum. To what degree do we suffer? Does an athlete who "suffers a loss" experience the same suffering as a family who "suffers a loss"? If I skip meals in an effort to lose weight because I suffer from being overweight, am I still suffering from hunger?

I meet many people who are deathly afraid of being stung by a bee. I'm not sure I would call getting a bee sting suffering. What about having only one shirt? Is that suffering? I guess that depends on how often you are able to do the laundry. What if you don't own a car and have to walk? That might depend on how far. What if you have to live in a one room shack? Well, that depends on who your room-mate is, doesn't it?

Interestingly, many people who live under poor conditions don't act as if they are suffering at all. They laugh, love, play, get married, have children, and seem to have a good, old time anyway. In fact, they don't seem to worry too much about suffering. They often seem happier than the rest of us.

I'm not sure why scientists worry so much about suffering. Suffering isn't a science, although my wife thinks I have turned it into such. She just doesn't understand how hard it is to be a professional windbag. Sometimes my back kills me from standing through all those lectures. I wonder if being boring counts as a disability. Do my students suffer?

If science is so interested in alleviating human suffering, why do almost all the things scientists discover to lessen suffering end up causing more suffering? After WW II, world scientists decided to eliminate Malaria. Things fell apart politically after a few years, and they gave up. By then, though, more people were dying from starvation because of overpopulation. Overpopulation was partly a result of more people surviving Malaria.

Scientists developed antibiotics to fight disease. But through the use of antibiotics, disease organisms became resistant to antibiotics. Wouldn't it be funny if we relied on pesticides to grow more food to feed the world, but the pesticides killed the bees, so we couldn't grow more food? No!

There are those who seem to be concerned about who is responsible for suffering. Interestingly, some people think it is God's fault for not stopping suffering from happening. Shouldn't they blame Satan for causing it to happen? If I fail to stop an accident, am I guilty of the suffering the accident causes? If a scientist fails to find a cure for cancer, are all the cancer cases scientists' faults?

Well, the truth is I became a scientist for several reasons. One was I discovered that I couldn't get a job with a degree in English literature. In addition, I liked working with things more than with people. I think I like humanity, as a whole, more than I like a lot of individual humans. Actually, I just have a lot of fun playing around with bugs.

Well, thankfully, I don't have to assign blame for suffering. If, like Dawkins, I never alleviate any suffering, I can always blame God.

NOVEMBER

Themes:
- Gratitude
- Thanksgiving
- Turning Colder
- Turkeys
- Turkey Dinner

Everything goes crazy about November. Halloween just got over and it's time to start the big plans for Thanksgiving: travel, dinner, company, family, church. But I am afraid that somehow or other the turkeys have taken over and dominate the month. That's sad, but not as sad as the turkeys.

NOVEMBER

Week 1

THE SCIENCE OF THANKSGIVING

Is turkey gravy a colloid or an emulsion? I've always wondered about that. But I only remember to think about it at Thanksgiving when I am too sleepy to think straight. Since turkey gravy has particles imbedded in a liquid, it must be a colloid. But since it has water suspended in oil, a liquid in a liquid, it must be an emulsion.

To avoid the question, we sometimes barbecue steak for Thanksgiving. I know it's not conventional, and for some it would be a disappointment. But having steak means I avoid having to make the hard decisions between dark and light meat. Hey, that's another thing. What makes meat dark and light anyway?

The color of meat comes from an oxygen-storing compound called myoglobin. Muscles have to have oxygen to provide energy. Oxygen is brought to the muscle by the blood which carries the oxygen with hemoglobin. There isn't any hemoglobin in the muscle cell, but the muscle cell has myoglobin. The oxygen carried in the blood is handed off to the myoglobin. Myoglobin makes the meat appear dark in color the way hemoglobin makes blood appear red.

Muscles that use a lot of energy require a lot of oxygen. The more oxygen required, the more myoglobin is needed. So the darker the muscle color, the more myoglobin there is in the meat. Turkeys use their leg muscles more than their wings, so their leg muscles have more myoglobin and are darker in color. Ducks, on the other hand, have dark meat for breast muscles because they fly more than they walk. White meat is white because it is just low on myoglobin.

Want to start an argument on Thanksgiving? Just strongly state your opinion on whether or not myoglobin enhances the taste of meat, or not, and see what happens. But if you ask people to "please pass the myoglobin", you will probably not get very much to eat. They probably won't know what you are talking about.

Myoglobin is a protein. These are very large, complex molecules. They are so big and complicated that, when they were first discovered as a class of compounds, early chemists called them by the technical term "glob". That's why a lot of proteins are called some kind of glob or another like myoglobin, hemoglobin, or maybe tetra-dihydro-benzylated-chickenwire-globin.

Turkeys have other suspicious compounds in them such as tryptophan. This is an amino acid that is also found in the human body. Humans can't make tryptophan, so they have to obtain it from their food. Tryptophan is used by the human body to make serotonin, a neurotransmitter that makes fruit flies sleepy. This has led some people to think that tryptophan makes humans sleepy also in spite of a complete lack of evidence that fruit flies celebrate Thanksgiving. The theory is that when we eat tryptophan, it causes our brains to make serotonin. The more serotonin there is, the more serotonin-activated neurons transmit. Serotonin-activated neurons are the ones that cause somnolence. It seems strange that more nerve transmission can make one feel like there is less nerve transmission.

But it may not be just tryptophan that makes you sleep through the second half of the game. Stretching the stomach and small intestine by eating a large meal causes the parasympathetic nervous system to reduce blood flow to your muscles and brain. The blood is shunted to the digestive tract which you have just overloaded. High protein and fat content in the meal tends to keep food in the stomach longer extending the period of time that the digestive system requires extra blood flow. That leaves less blood to be delivered to the brain. Neurons do not possess myoglobin to store oxygen. So when blood flow declines, nerve activity slows. It is not a profitable time to contemplate whether or not turkey gravy is a colloid or an emulsion, no matter how pressing it may seem.

NOVEMBER

Week 2

FEATHERS

Do you know how many feathers a turkey, *Meleagris gallopavo*, has? I don't either. Apparently no one does. Can you believe that? You'd think we would. Ben Franklin thought the turkey should be our national bird. I wonder if we know how many feathers the Bald Eagle, H*aliaeetus leucocephalus*, has.

The bird with the least number of feathers is the *Archilochus colubris* with only 940. Okay, I didn't actually count them myself. Someone must have though because that's what it says in my bird book. Ornithologists don't lie about important things like that. I just don't know why they would know how many feathers a stupid hummingbird has, but not a turkey.

Archilochus colubris is commonly called the Ruby-throated Hummingbird, and they are only about eight centimeters long and weigh about three grams. It would take about 4500 hummingbirds to make up one turkey. Be thankful they aren't served for dinner, I guess. We don't have them in Colorado. They only nest in the eastern US in the summer time. In the winter, they sensibly head down to Central America. I'm not sure how they get there. Maybe they have frequent flier miles.

The Whistling Swan, *Cygnus columbianus*, can have as many as 25,000 feathers during the winter. The number isn't the same in summer because they molt their feathers on a seasonal basis. Swans are fifteen times bigger than Ruby-throated Hummingbirds, but have twenty five times as many feathers. What's that all about?

Birders are pretty serious about what they do, but I am not sure if I really believe these numbers. I kind of suspect someone cheated. Since about a third of a bird's feathers are found on its head, I'll bet they just counted the feathers on the head and multiplied by three. Besides, I have plucked chickens, and I don't believe anyone can even count all those feathers.

The feathers that we see on the outside of the bird are called tetrices. They are the feathers that shape the body, and add the ruby color to the throats of some hummingbirds. Just the male birds have these ruby throats. Maybe ruby throats are to male hummingbirds what big biceps are to human males. Tail and wing feathers are called quills. Then it gets confusing when you get "down" to the underwear. Sorry! Soft feathers, called plumules, help trap and hold heat. Even tinier, hair-like feathers under these are called filaplumes. Do you seriously believe someone counted twenty-five thousand of these various feathers on a swan?

Back to turkey feathers though, part of the problem is that turkeys are changing size. The average turkey weighed about 13 pounds in 1929. Today the average turkey weighs 30 pounds. That's even bigger than a *Cygnus columbianus!* Since the turkey is 4500 times bigger than a ruby throated humming bird, perhaps they have 4500 times as many feathers. That would give them 40860 feathers.

About four billion pounds of feathers are produced every year by the poultry industry in the United States. About 140 million turkeys are killed and about nine billion chickens, *Gallus gallus domesticus*. Remember though, turkeys are three to four times bigger than chickens which only average about eight pounds. So we could say that about a billion pounds of feathers come from 140 million turkeys every year, and that's about seven pounds of feathers per turkey.

Feathers vary in weight depending on the type of bird they come from and where the feathers come from on the bird. To make one pound of ruby-throated hummingbird feathers can easily take thousands of feathers. To make one pound of large feathers from an emu or ostrich can take around 200-500. If we estimate 1000 feathers per pound, then turkeys have, on the average, 7000 feathers.

So turkeys have between 7000 and 40,000 feathers depending on how you calculate it. You heard it here first. Why do I always have to figure these things out myself?

NOVEMBER

Week 3

DEPRESSED TURKEYS

There are a lot of depressed turkeys these days. It's not hard to tell a depressed turkey from one that's in good spirits. Depressed birds stand with their heads tilted downward or drawn into their bodies. Their feathers are ruffled, and their wings droop. Their eyes are partly closed. The birds may act alert and agitated when disturbed, but they quickly become lethargic again. By comparison, normal turkeys are highly excitable and often seen strutting around as if they had good sense, which they don't.

It's hard to determine the cause of turkey depression. It's not that they don't have adequate reasons for being depressed. More than 140 million of them die each year in the United States alone! Even if they, themselves, don't meet an untimely fate, they have undoubtedly lost friends and relatives, or soon will. But it is a far more complex matter than it first appears.

Turkey depression seems to run highest in urban-turkey settings such as turkey farms and large cities, where feelings of isolation, in the midst of crowds, are a common phenomenon. In addition, crowding seems to bring out unusual behaviors in turkeys, including alienation from authority, violence, mob rule, sexual deviation, and public health risks that contribute to poor mental health. Feelings of dependency and worthlessness aggravate the condition.

A number of names have been proposed to describe the condition that is turkey depression: turkey blues, occupiensis, or blackhead, are common. But as with humans, the causes of turkey depression are several and careful diagnosis is required.

Actually there are three distinct conditions that scientists have categorized. The first, "turkey blues" may be nothing more than serotonin imbalance. The timing is highly seasonal and may be related to day length, a condition commonly called Seasonal Affective Disorder (SAD). The second condition, known as occupiensis, may affect as much as ninety-nine percent of the turkey population. The severity of symptoms these turkeys exhibit apparently vary exceedingly.

Blackhead, on the other hand, is an actual disease that is well known among turkeys and other bird brains. It is caused by a parasite called *Histomonas meleagridis.* This single-celled parasite affects several bird species, but it seems to be most severe in turkeys. It is also known as infectious enterohepatitis, or histomoniasis. The parasite inhabits the birds' intestinal tract but also invades the liver. In both places, it causes extensive tissue degradation. That's what is so depressing.

Adding insult to injury, *Histomonas* is transmitted between turkeys by another parasite, *Heterakis gallinarum,* which is an intestinal worm of turkeys. So . . . the parasite of turkeys is transmitted to turkeys by parasitizing a parasite of turkeys. This is reminiscent of the old nursery rhyme:
Big fleas have little fleas
Upon their backs to bite 'em.
And little fleas have lesser fleas,
And so, ad infinitum.

The Blackhead organism cannot survive outside the turkey very long, but it can become encapsulated in the egg of the *Heterakis* worm. The egg of the worm can then survive for many weeks depending on environmental conditions.

Turkeys with advanced Blackhead develop a sulfuric diarrhea that spreads the eggs, and *Histomonas,* over a wide area. Another slight twist to the whole situation is that earthworms often ingest the *Heterakis* eggs and carry them deeper into the soil. Here the parasite survives longer and is spread out over an even larger area.

Make no mistake, though. Turkey depression costs you a lot of money. Mortality from turkey depression can be as high as eighty to ninety percent, and surviving turkeys are seldom worth much. The cost of depressed turkeys must be compensated for by the price of surviving turkeys. Consequently, depressed turkeys cost us money.

Oh, and the increased cost for the un-depressed turkeys leads to even more depression, at least among turkey consumers. This year there seems to be more than the normal number of depressed turkeys out there. It's very depressing.

NOVEMBER

Week 4

YAWN

YAWN! Nice nap? After Thanksgiving dinner, I mean. This holiday got me to wondering about why we yawn. And guess what? We don't know. Yep, the whole yawning thing appears to be a big mystery to scientists. Wow! How can we not know the reason for yawning? And people think science is boring.

So I decided to demonstrate just how daring a scientist can be and devote an entire chapter to yawning, in spite of the evidence that yawning is contagious. There is a huge risk of putting my readers to sleep by the end of the fourth paragraph. Live dangerously, I always say.

The study of yawning is called chasmology. (I'm not making this up.) A yawn is an ancient, stereotyped, involuntary reflex involving stretching the body. There is the opening of the mouth wide, followed by a slow inspiration of air, stretching of the ear drums, expansion of the lungs, increased heart rate, and the rapid exhalation of air. It has been shown that all reptiles, birds, and mammals yawn. YAWN!

There are a lot of theories about why we yawn, but most of them have larger chasms than the yawn itself. For example, it is commonly thought that we yawn to get extra oxygen, perhaps because we haven't been breathing deeply enough. However, Robert Provine from the University of Maryland showed that neither increased oxygen nor decreased carbon dioxide altered the yawning rate. Actually, yawning slightly decreases blood oxygen.

Another theory is that yawning is some ancient, visual, communication system like smiling or frowning. (Are you still awake?) Maybe it was originally a polite way of saying "I'm tired. Go home now". I don't think this is a very good explanation. It is hard to imagine why a reptile would need to yawn to tell the other snakes to go away.

Maybe we yawn to stretch our lungs. If they get a little wrinkled, like balloons that have become deflated, a yawn would stretch them out again. But why would you need to stretch out your lungs before going to sleep? And is there any evidence that lungs get wrinkled anyway?

A lot of people just think yawning means you are tired or bored. So why do people yawn early in athletic activity? Besides, I reject that theory because all those students in my class couldn't possibly be that bored. YAWN! Besides, people yawn when they wake up also. Are we bored of sleeping? In fact, some studies suggest that humans yawn more upon awaking than they do when somnolent. If morning yawning is to wake up, what is evening yawning for?

The most recent theory is that we yawn to cool our brains. Extended brain usage overheats the brain. When we draw fresh air in, it cools the blood and speeds circulation to the brain. Gordon Gallup, a psychologist at University of Albany, did an experiment where he asked students to watch videos of people yawning. (By the way, are you still awake?) If the students breathed through their mouths and held hot compresses to their head, they yawned more than if they breathed through their noses, a natural brain coolant activity, and held cool compresses to their foreheads. I'm pretty sure this theory isn't true both because I yawn a lot, and my brain is not so hot.

One of the neat things about science is that when we discover that we don't know something, it is very freeing. Since no one knows, we can imagine whatever we want until someone proves us wrong. So here's my theory. I think we yawn when we change states of activity. As we drift to somnolence, perhaps we yawn to step down the oxygen levels. Then as we awaken, we yawn to increase blood flow to muscle and brain. YAWN! I think that when . . . YAWN! . . . ever we . . . YAWN! . . . .

NOVEMBER

Week 5
(left overs)

IT'S GETTING COLD

On the way to work the other day, I saw a squirrel burying Sterno. I think it's going to be a long winter.

But at least we know how cold it is. That hasn't always been the case. In fact, people didn't know how cold it was until 1724, and even then they weren't sure. That was the year Daniel Gabriel Fahrenheit published a paper proposing the Fahrenheit temperature scale. Prior to having an actual reproducible scale, people had to tell the temperature simply be referencing how cold various body parts were on people engaged in known occupations such as well diggers and witches.

Fahrenheit invented his scale as part of a diabolical plot to get rich. While he was a German engineer, he was also a glass blower. He happened upon the idea of making glass tubing filled with various liquids to measure how much the liquid contracted or expanded when hot and cold. He thus invented both the alcohol and mercury thermometers. After establishing his scale, he was able to open a lucrative business selling his handmade thermometers.

Contrast this to the many musicians who labored over the centuries to create musical scales and never received a dime. It's rumored that Mozart was once so cold that he had to chop up his piano for firewood. However, it only gave him two chords.

Fahrenheit arrived at his measurements simply by comparing measurements in three fixed references. First he made a mixture of ice, water, and ammonium chloride.

Placing his newly invented thermometer in it as it reached equilibrium, he marked the lowest point and called it 0° F. Then he used the same crude thermometer to mark the point in plain water just as it was freezing. Then he measured the point on the thermometer, taken under the arm, for body temperature.

Later work by others showed that water boiled at 180°F above freezing, so the scale was adjusted slightly to make that number standard because it is easily divisible by many other numbers. That established 32°F as freezing and boiling water at 212°F. Fahrenheit said body temperature was 96°F, but because of the adjustment we now call normal body temperature 98.6°F.

Just how cold is it? It's so cold I heard that Al Gore returned his Nobel Prize.

However, just as people were beginning to think they knew how cold it was, Anders Celsius became interested in temperature. One morning in Sweden, it was so cold that he saw a politician with his hands in his own pockets. Fascinated, in 1742, he immediately set to work performing some experiments. In these experiments he showed that the freezing point of water was independent of latitude (or barometric pressure), but the boiling point varied with barometric pressure.

He proposed a new scale called the centigrade scale, meaning one hundred steps. Since water freezes at the same temperature at all latitudes, he made the freezing point of water 0°C. Then to maintain the water standard, he declared boiling water at sea level as 100°C. But he got a little confused and called it 0°C for boiling water and 100°C freezing. No I didn't make a mistake. That was his original idea. I told you he was from Sweden.

But now the world had two thermometers: one of which was upside down. People have never been the same since.

Most of the world uses the Celsius scale, but the US and Belize still use the Fahrenheit scale for daily use. In 1745, just a year after Celsius died, Carl Linnaeus had the good sense to turn the Fahrenheit scale over to facilitate calculations. I'm not sure who gave him permission to do that, but I'm glad he did.

Anyway, it sure is cold. It's so cold that on the way home today I saw a hitchhiker holding up a picture of a thumb. At least, I think it was a thumb.

DECEMBER

Themes:
  Cold
  Holidays
  Frostbite
  Christmas
  Christianity

    December is such a special month, even for those who aren't Christians. There is the Pagan holiday of the New Year and the Jewish celebration of Hanukkah. December marks the ski season for the winter sports aficionados. Santa Clause and his reindeer ride on the 25th of the month. Plus December marks the winter solstice. I'm not sure I can fit it all in.

DECEMBER

Week 1

HOLE-Y SNOW BATMAN, IT'S AN AIRPLANE

Here in Colorado, I have watched a rain storm across the street leave me completely dry. It can rain heavily in one part of the valley, but not in another. Like my Dad always said, "If you don't like the weather, wait ten minutes."

But in December of 2007, there was apparently a very unusual snow storm that took place on the front range of Colorado not far from Denver International Airport (DIA). It lasted for about forty five minutes. The storm was about twenty miles long, but only two and a half miles wide. It covered the ground with about two inches of snow. What could cause such a long, skinny snowstorm? Apparently it was caused by hole-y clouds.

Snow is made of water. Water molecules are arranged in the shape of a "V". The oxygen is at the point of the "V" and the two hydrogen atoms are at the ends of the arms. The three atoms do not share electrons equally. The oxygen has all the electrons most of the time, and that makes the oxygen have a net negative charge. The hydrogen atoms seldom have the electrons, so they are mostly positive. This makes the water molecule have a slight polarity: a positive end and a negative end. Since the situation is sometimes reversed momentarily, this is called a dipole moment (two poles changing from moment to moment).

But this means that water tends to cling to itself with the positive ends being attracted to the negative end, and vice versa. As the water molecules cool they settle into position aligned positive to negative and get closer and closer together. At about four degrees Centigrade they are packed as close as they can get, and water is as dense as it can get.

Water typically freezes at zero degrees Celsius. During freezing, water actually expands slightly and become a little less dense, which is why ice floats.

But water clings to itself and is so small that it can sometimes cool below freezing without forming an ice crystal. It needs a surface to adhere to in order to form the crystal. Hence water can remain liquid sometimes at temperatures as low as - 40 C. However, in that super cooled state, even slight disturbances can cause the molecules to suddenly crystallize. If there is a dust particle, or another ice crystal, or sometimes even air movement, pushing the molecules together can cause the formation of ice crystals.

Ice crystals that form at a high altitude can fall into a cloud below that has super-cooled water, causing condensation of the water in that cloud and leaving behind a hole in the cloud filled with water crystals. These are called hole-punch clouds.

It turns out that Andrew Heymsfield, an atmospheric scientist at the National Center for Atmospheric Research in Boulder, discovered our freak, 2.5 mile wide snowstorm. It happened because just such a hole-punch cloud created by the movement of a turboprop jet through a super cooled cloud on its approach to DIA. While reviewing images taken from a plane-mounted camera, they noticed a long thin hole in an otherwise solid layer of clouds. Beneath this hole, a wall of falling snow extended from the hole all the way to the ground. The hole lined up perfectly with the flight path for planes approaching DIA on that date.

As the moisture-rich air passed the front facing edge of the planes propellers, the air accelerated, its pressure decreased, and the temperature dropped about 10 C. Ice crystals were formed. These sucked up surrounding cloud water to gain enough weight to fall through more super-cooled air below triggering a snow storm below the plane trajectory.

    Cloud seeding has been used for a long time. But apparently airplanes, alone, can substantially increase precipitation as they fly through super-saturated clouds. Maybe THAT's why the weather changes so frequently in Colorado. Too many airplanes?

DECEMBER

Week 2

THE COST OF THE HOLIDAY

Scientists routinely come up with new ideas for products, procedures, and technologies. But they almost never acknowledge the true costs of production. Of course they calculate prices, the prices being what the products will sell for. Strangely, though, the prices seldom reflect actual costs.

Let me use scientists' relatively-new-found ability to manipulate genetic traits as an example. Imagine that we can use genetic engineering to make a new milk cow that can produce 50,000 pounds of milk a year. Great idea! But such a cow would also have greater energy demands. That kind of productivity would require at least a bushel of grain a day just to make the milk, let alone grow the cow. The 365 bushels of grain per year for such a cow would be obtained through competition with all other grain users. The demand for grain would cause the cost of grain to go up.

The increased prices for grain would cause an increase in the number of farmers growing grain. To grow more grain would then require more acreage under grain cultivation. More acreage would require more, and perhaps bigger, machinery. Competition for machinery would drive up the cost of machines.

More and bigger machines would mean an increase in the use of fuel, driving up the costs of petroleum products for everyone. The extra costs of fuel would necessarily increase the costs of grain production and increase the cost of grain for all users, the dairy farmer and the consumer.

Modern agriculture is mostly mono-cropping, and this practice causes soils to be seriously depleted of nutrients. The increased demand for grain would further deplete the soil through compaction, erosion, and over production.

Depleted soils require more fertilizer. Fertilizer is often manufactured from petroleum products. Even the use of natural fertilizers requires fuel for extraction, processing and delivery, further driving up the cost of fuel.

The demand for genetically-engineered cows would increase as competition grows between dairy farmers. Yet these cows would be rare and expensive to obtain. Expensive cows require further expenditures such as special barns, veterinary care, special milking protocols, and insurance policies to safeguard the investments. All of these things would increase the costs of business to the farmers, and they would pass all these added costs on to the dairymen and the consumer.

The increased supply of milk could decrease the demand for dairy farmers, and some would go out of business. The out of business farmers might then move to the city. The city would raise taxes to pay for the additional services required by a growing population.

In the meantime, the price of milk might actually go up from all the increased expenses and the decline in the number of producers. If milk prices dropped due to excess milk production, the government, that is the tax payers, would pay milk subsidies to the remaining dairy farmers to not produce milk with their genetically enhanced cows.

In the end, there is almost no way of knowing whether the scientists would have created more and cheaper milk, or a tremendous economic morass that results in less, or more expensive, milk. It is obvious that two things in the list of issues are in short supply and are maybe even irreplaceable: energy and top soil. Those costs are never considered by scientists.

This holiday season there are new telephones, electronic

books, toys, games, home theaters, tools, and new fashions and fabrics. The production of these products are based on the bedrock of science and technology. For the most part, they are all reasonably "priced" and on sale for the holiday. But do we know what they "cost"? We used to only have to worry about our own Christmas indulgences. Should we now be considering the cost of Christmas to us and everyone else? The retail stores are counting on us spend a lot this year, to lift the economy. I'm just not sure we can afford the cost.

DECEMBER

Week 3

COUNTER CURRENT CHRISTMAS

Have you ever worried about reindeer getting frostbite? I used to think about this a lot, which tells you something about my personal life. It is confusing, because when I watch National Geographic specials, the reindeer are all just out there in the open. It gets pretty cold up at the North Pole. Maybe Santa keeps his reindeer in a special barn or something. But still, flying around for hours on Christmas Eve must put them at risk for hypothermia.

Ever since I went to school and learned biology I don't worry about it anymore. Knowledge can be a wonderful thing. Education isn't just about having a fatter paycheck. No, there is peace of mind, understanding, clear thinking, and finer sensitivity to be had. That and the opportunity to remain out of the work force for several more years.

Reindeer are especially designed for their environment in a most unique way. Most of a mammal's body heat is found in their body mass. Appendages have smaller surface areas per volume, so they tend to radiate heat. That is why, when you exercise in the heat, your hands tend to swell. The blood is shunted out to the body's periphery such as fingers, toes, ears and nose to radiate excess heat.

In the winter, the opposite can happen. The body tries to conserve heat by keeping it in the core of the body and slows the circulation to the extremities. That is why your rings slip off your hands when they are cold. There is less blood flowing into the fingers. That is also why your fingers, toes, ears and nose are the first to get cold, and sometimes frostbitten. The body decides to lose them instead of your core heat.

But reindeer don't lose their rings in the winter because of the way the animals' circulatory system is structured.

For example, water runs across fish gills in one direction and the blood pumps through in the opposite direction making for an efficient exchange of oxygen and carbon dioxide. In our kidneys, the forming urine flows one direction in one tube and in the opposite direction through another. This process helps extract sodium across the membranes and concentrate it in the tissues.

But the big thing is that the arterial blood flows down into the reindeer's feet on the outside and the venous blood returns to the body on the inside. However, the two are located exactly adjacent to each other, just the blood moving in opposite directions. Because the circulatory system is insulated with fat and hair, the heat in the warm arterial blood is given off to the already cooled venous blood and returned to the core of the body. This helps them avoid hypothermia and still deliver warm blood to the extremities.

This Christmas eve you can rest easy concerning Santa's reindeer outside in the cold knowing that nature has provided them with an efficient counter current mechanism for retaining heat in the face of the North Pole winter.

DECEMBER

Week 4

CHRISTMAS

There are many human experiences that are enhanced by scientific knowledge, and I celebrate them in this book. But knowing that the earth's axial tilt is 23.26° does not make the winter more beautiful or Christmas sweeter. The uneasy hearts and fears accompanying dim days and dark nights are seldom mitigated by knowledge of why it is dark, and that it will pass. The joys of warmth and firelight are seldom enhanced by material circumstances and facts.

No, there are realms of our existence in which science has very little to offer. When disaster strikes and hopes and hearts are crushed, science really has little to say. Our knowledge may help those who remain to live and rebuild their lives, but it has little to give in the way of comfort. Also, when hearts are full, science does not increase the sweet emotions we feel. Science serves mankind in a material way, but often leaves us wanting more. Perhaps that is why so many scientists dabble in the arts.

However, there is much to say in defense of science. And other fields that offer solace to the soul often leave men suffering from the material world from which science offers relief. Grief is compounded by physical suffering, and science has greatly served mankind's physical wellbeing. So I have struggled to find an appropriate scientific message for this special time of year. Here is what I came up with.

It is amazing the way living things tenaciously hang onto life during cold and dark times. Our experience teaches us that everything is not dead, but that life waits to begin again. If life is sacred, then the DNA code is sacred. And there is no more purified and refined form of DNA in nature than that which is found within a seed. Seeds are the basis of life! They have become the predominant metaphor for creativity, inspiration, faith, renewal and resurrection. In the dead of winter, seeds are the dead of winter.*

Implicit in the Christian story for the last two thousand years is the birth, life, death, and subsequent resurrection of Jesus Christ. This is the same story celebrated by ancient traditions surrounding the winter solstice. Both stories are entwined with ideas of death and resurrection, the cycle of the seasons, dark giving way to light, repentance, growth, and change. Christ's story reminds us that there is light in the darkest hour. But it also speaks to us of darkness and danger even as we rest securely in the cradle. So whether we are Christian, pagan, Jew, or completely irreligious, this season can speak to our hearts in a way that science fails to do.

*THE DEAD OF WINTER
The leaves have fallen from the trees
The trees don't need them anymore
Colder now it's plain to see
The ground needs a blanket more
The dead of winter is at the door
The seeds of harvest have all been stored
Each awaits something more
To live again just like before

Flowers bow weary heads down
And give their children to the earth
The final petals come unbound
Giving birth to all they're worth
The dead of winter are at the door
The seeds of harvest have all been stored
Earth and stone becomes the door
Through which we pass to life restored

Wild seeds fall upon the ground
And await the coming Son (sun)

Seeds are laid in funeral mounds
To await resurrection
The dead of winter are at the door
The seeds of harvest have all been stored
This is what the Son (sun) is for
Restoring life, the earth's savior

DECEMBER

Week 5

CHRISTIAN SCIENCE

This thing called science has only "happened" once in the entire history of human endeavor. Doesn't that thought seem strange? There have been many great nations and rulers. There have been many passing governments and cultures. But, only once in history, did science come about.

So where did science get its start? When? Why at that time and place? Why did science not develop under other circumstances? You may be surprised at what follows. Science was born of religion. In fact, science was born of Christianity. I think that specific understanding might be especially significant at this time of year.

Early man was ruled by animism, a philosophy that gave every rock, tree and animal a spirit. The world was mysterious and unpredictable. What followed were numerous polytheistic religions such as those practiced by the Babylonians, Egyptians and Greeks. While each culture contributed additional ideas, none of them produced the ability to predict outcomes and control the world in the way modern science has provided.

Eventually three monotheistic religions developed: Judaism, Islam and Christianity. If I understand them correctly, both Judaism and Islam proclaim there is a divine law-giver and a book of laws. In the case of the Jews, these laws apply just to the Jewish people. For the followers of Islam their book of laws applies to all mankind. Jews and Muslims debate amongst themselves as to the best ways to interpret and apply their books of laws.

Christianity has no book of law. Instead, Christianity is based on doctrine, or a set of beliefs, about man's relationship with God. The New Testament is not a book of laws. It is the story of the Son of God, a man named Jesus Christ, who taught that each individual should have a personal relationship with God.

But exactly what should this relationship be? I suspect Christians may argue amongst themselves, more than Jews or Muslims, because that relationship is not spelled out in a book of laws. Instead, it has been left up to the individual to determine what that relationship should be. A long list of Christian thinkers wrestled with various theological problems: Paul, Tertullian, Clement, Augustine, Thomas Aquinas, Luther, and many more. Note here, though, that it has always been acceptable in Christianity to apply reason to understanding God.

It was the acceptance of "reason", as a way of knowing God's mind, which led to the medieval establishment of monasteries, schools and universities. Here the first agricultural, medical, and astronomical studies were conducted. Early Christians believed that by understanding nature, we would learn more about God

Science was born in the fourteenth century out of a dispute between the following two Christian ideas. On one hand, some theologians believed that the nature of God could best be discerned through scholarly debate and reason. Other theologians came to believe that experience with the real world, what we might now call experiments, were the best way to determine the nature of God. Later the two approaches melded into our modern study called science.

Religion is sometimes portrayed as "lacking in reason". Yet it is the acceptance of reason as a mode of inquiry, a process called deduction, when combined with the acceptance of experimental evidence, a process called induction, which led to the creation of science as a method for learning about the world. The majority of early scientists were Christians, who saw themselves as theologians trying to determine the nature of God through the use of reason when applied to the study of nature.

This is not to say that there are not brilliant Jewish or Muslim scientists only that those cultures did not give rise to the discipline of science itself. In the eighteenth century, science developed a schism and began to be more secular. Since then, there have been both scientists and theologians who have declared some kind of academic or spiritual war between the two subjects.

But this Christmas season, perhaps we should recall that science was born this day, in the City of David, two thousand years ago.

EPILOGUE

Scientists talk about things differently than normal people. I don't think it's intentional. Maybe the same sort of thing happens in other disciplines. After a while, language simply starts to mean different things to different people based upon their shared experiences.

The reason we are able to communicate at all is because we share some similar experiences. We think the sky is blue because we were told it is blue. If someone had told you as a toddler that the sky was red, like someone told me (you know who you are), growing up normal like I have would be a miracle. When I say that a destination is a block away, most of us have some shared experience with what a block means.

However, when we make the decision to delve deeply into any subject, the experiences we have compared to the experience of others can be different. At that point language can start to have different meaning from that of normal people. (I am assuming here that only abnormal people delve deeply into subjects.) A good example of this is watching ESPN. What are those guys talking about?

This column is a good example of the strange use of language. I maintain that I am normal despite harboring earlier beliefs that the sky is red. I then say that only abnormal people delve deeply into subjects, which I have sometimes done. My saving grace, as far as being normal is concerned, (assuming I have any saving graces) is that I think I might have ADD and seldom delve deeply into anything before I get distracted.

Now, what was I saying? Oh yes, scientific language. It goes beyond just the use of weird Latin terms like "*Macrocanthorhynchus hirudinaceous*", or chemical jargon like "six hydroscopic ambaphascient tetra halide". Sometimes understood concepts in science become obscured in everyday usage.

For example, most people know that when they speak about "neutrality" they are talking about something in the middle between two extremes. In fact, that may not be what "neutral" means at all for scientists. For example, in chemistry, a neutral solution is a solution that is neither acid nor base. Neutrality is sometimes achieved by mixing acids and bases in such a manner that they cancel each other out. However, it can also be achieved in a solution that simply does not ionize. In other words, a neutral solution is not in a state of being acid or base.

In physics, neutral refers to the lack of any electrical charge. Neutral is not half positive and half negative. It is non-electric, at least at the moment. In the physical world, a car in neutral is not half way between fast and slow, or forward and back. A car in neutral isn't in any gear at all.

I guess most of the time these little differences don't matter. If I don't know that the "five hole" is the gap between the goalies' knees in hockey it doesn't make much difference. I don't watch hockey. But sometimes the nuances of language can be worth examining.

When we use the word neutral to mean something or someone in the middle, it may be very misleading. What did it mean to not take sides in World War II? Switzerland was neutral. Did that mean they were in the middle between good and evil, that they aren't in either condition, or that they don't know the difference? Since good and evil depend upon a standard, like gears or electrical currents, does that mean Switzerland had no standards of good and evil?

What does it mean when people maintain that government has to be neutral when it comes to standards of behavior such as are found in religion, ethics, or morality? Does that mean government must take no stand, or does that mean there are no standards for government? What if the sky really is red and we've all been told wrong?

So be kind. This book is neutral when it comes to science.

www.ingramcontent.com/pod-product-compliance
Lightning Source LLC
Chambersburg PA
CBHW051908170526
45168CB00001B/287